Electricity Market Reforms

Lev S. Belyaev

Electricity Market Reforms

Economics and Policy Challenges

Springer

Lev S. Belyaev
Siberian Branch, Energy Systems Institute
Russian Academy of Sciences
664033 Irkutsk, Russia
belyaev@isem.sei.irk.ru

ISBN 978-1-4419-5611-8 e-ISBN 978-1-4419-5612-5
DOI 10.1007/978-1-4419-5612-5
Springer New York Dordrecht Heidelberg London

Library of Congress Control Number: 2010935847

Printed on acid-free paper

Springer is part of Springer Science+Business Media (www.springer.com)

Acknowledgement

The author is very grateful to the corresponding member of RAS N.I. Voropai, to Professors A.P. Chernikov, I.I. Golub, L.P. Padalko, S.I. Palamarchuk, B.D. Syutkin, and V.I. Zorkaltsev for reading the manuscript and making valuable remarks that facilitated its improvement, and also to colleagues Doctors L.Yu. Chudinova, V.V. Khudyakov, O.V. Marchenko, S.V. Podkovalnikov, V.A. Saveliev, and G.B. Slavin, whose joint works contributed to the writing of the book. Special thanks are extended to Professors Ferdinand Banks, Dorel Soares Ramos, Hugh Rudnick, Steven Stoft, and Dr. Marcel Lamoureux for the presented materials and consultations on the electricity markets in the USA, South America, and Western Europe. The author expresses an exclusive gratitude to L.K. Rogova, O.M. Kovetskaya, and E.G. Lapteva for preparation of the manuscript for publication and to V.P. Ermakova, M.V. Ozerova, and N.V. Zhitova for its translation into English.

Contents

Abbreviations

AC	alternating current
AO-Electrostantsiya	joint stock company of a power plant (in Russia)
AO-Energo	joint stock company of regional EPS (in Russia)
BETTA	British Electricity Trading and Transmission Arrangements
BM	balancing market
CCI	combined-cycle installation (on natural gas)
CCPP	combined-cycle power plant (with CCI)
CGPP	cogeneration power plant
CIS	Commonwealth of Independent States (of the former USSR)
cop	Russian copeck (0.01 ruble)
CPP	condensing power plant
CRF	capital recovery factor
CRFEG	capital recovery factor at expending generation
DAM	day-ahead market
DC	direct current
DSC	distribution–sales company
EDF	Electricite de France
EPS	electric power system
ERI	Energy Research Institute (the Russian Academy of Sciences)
ESI SB RAS	Energy Systems Institute of Siberian branch of Russian Academy of Sciences
ESR	Energy Strategy of Russia (till 2020)
EU	European Union
EUPS	European section of UPS (in Russia)
FACTS	flexible alternating current transmission system
FEC	Federal Energy Commission (in Russia)
FERC	Federal Energy Regulation Commission (in the USA)
FNC	Federal Network Company (in Russia)
FOREM	federal wholesale electricity and capacity market (in Russia)

gce	gram of coal equivalent (7.0 kcal or 29.3 kj)
GTI	gas turbine installation (simple-cycle)
GTP	gas turbine plant (with GTI)
HPP	hydropower plant
IEF	Institute of Economic Forecasting (the Russian Academy of Sciences)
IPP	independent power producer
IPS	interconnected electric power system
ISEPI	interstate electric power interconnection
ISET	intersystem electric tie or interstate electric tie
KEI	Krzhizhanovsky Energy Institute (in Russia)
LES	large energy system
NETA	New Electricity Trading Arrangements (in Great Britain)
NNC	national network company
NORDEL	electricity market of Scandinavian countries
NOREM	new model of wholesale electricity and capacity market (in Russia)
NPP	nuclear power plant or new power producer
NPS	National Electric Power System
PGC	power generation company
PJM	market in the states of Pennsylvania, New Jersey, and Maryland
PRDS	pressure-reducing and desuperheating station
PSPP	pumped storage power plant
RAO	Russian joint stock company
REC	regional energy commission
RES	renewable energy source
RF	Russian Federation
SC	sales company
SO	System Operator
STI	steam turbine installation
tce	ton of coal equivalent (7.0 Gcal or 29.3 Gj)
TGC	territorial generation company (in Russia)
TL	transmission line
TNC	transport network company
TPP	thermal power plant (on fossil fuel)
TSA	Trading System Administrator
UPS	Unified Electric Power System (of Russia)
VIC	vertically integrated company
WGC	wholesale generation company (in Russia)
WTO	World Trade Organization

Microeconomics Terms

AFC	average fixed costs
ATC	average total costs

AVC	average variable costs
D	demand
FC	fixed costs
HAVC	hourly average variable costs
HMC	hourly marginal costs
HVC	hourly variable costs
LAC	long-run average costs
LMC	long-run marginal costs
MC	marginal costs
MR	marginal revenue
S	supply
SATC	short-run average total costs
TC	total costs
VC	variable costs

List of Boxes

Chapter 1
Introduction

Our life is rich in various phenomena. Some of them are amazing and hard to explain; moreover, sometimes they oppose common sense, theory, and practice. A phenomenon of the kind has been observed over the past ten or more years in the electric power industry of many countries, including Russia. Incidentally, the fact that this does not happen in all countries is significant and optimistic. This book is devoted to the analysis of this situation, an attempt to understand its roots, trace its consequences, and find the ways of prevention.

It concerns the reform (restructuring, deregulation, liberalization) of the electric power industry with the transition from vertically integrated regulated monopoly companies to *a competitive market* (with unregulated prices). Russia started the transition in 2001 after the Decree of the RF Government No. 526 had been passed [1].

Transition to a competitive market poses many questions that can hardly be answered convincingly:

For example, "Why should the industry, which enjoys economies of scale and is, therefore, a natural monopoly, be forcedly split into many companies and deprived of this benefit?"

Or, "Why should the buyers pay for the products of cheap generators (for example, hydro power plants) the price of the most expensive (marginal) generator that has to be involved to meet the demand of consumers?"

Or another question, "If new generators should be paid high prices for their products to pay back their investments, then why pay the same high prices to the operating generators?" and so on.

This book addresses the electric power industry in countries with market economy. During planned economy in the USSR the theory and practice of managing the development and operation of the industry were worked out profoundly. The problems discussed in the book simply did not exist.

The terms "reform" and "restructuring" will be used as synonyms in a wide sense to describe any structural transformations in the electric power industry, and the terms "deregulation" and "liberalization" will be used (also as synonyms) only in the sense of termination of the governmental (or regional and municipal) electricity

L. S. Belyaev, *Electricity Market Reforms*,
DOI 10.1007/978-1-4419-5612-5_1, © Springer Science+Business Media, LLC 2011

price regulation. This difference is quite important to characterize the process of reforms.

The causes of reforms in the electric power industry and the goals posed are country-specific. In *developing* countries reforms were a result of insufficient governmental funds to ensure the required power development, and the main goal, therefore, was to attract private (including foreign) investments. Some countries, however (for example, China and India), retained the regulation of electricity prices, as their liberation under the conditions of power shortage was just impossible. These countries did not deregulate the industry, i.e., did not make a transition to a competitive market. At the same time some other countries (for example, Chile, Argentina, and Brazil) created competitive wholesale electricity markets.

In the majority of *developed* countries the main cause of reforms was high electricity prices, and the reforms aimed to decrease them. Competition in electricity generation and sales was expected to enhance the efficiency and decrease production costs and, hence, the prices for the final consumers. Many developed countries (England, some states in the USA, Australia, and Scandinavian countries) have deregulated their power industries and organized competitive wholesale and retail markets with free prices.

Meanwhile, the experience of the past years [2–12] shows that electricity deregulation (or liberalization) often leads to the opposite results, i.e., to a price rise, lack of investments, power shortage, and decrease in electricity supply reliability (including blackouts). The initial concepts of reforms are revised (reform of the reforms), the process of reforms is delayed (none of the countries has completed reforms), electricity markets grow more complicated, the proposals are put forward to restore regulation, etc.

The main goal of the book is to show a general imperfection of the electricity market, the flaws of competitive market, and the necessity (inevitability) of state electricity price regulation. The content of the book is based on the author's long-term research of the management of operation and expansion of the Soviet Union electric power systems (EPSs) [13–16], on his own studies of electricity markets [17–22], and on the analysis of the world's experience.

Electricity market imperfection (in terms of microeconomic theory) is caused by the properties of EPSs. It is commonplace to speak of the electric power industry restructuring; in fact, however, we are dealing with the restructuring of very complicated technologically interrelated EPSs. With transition to a competitive market, EPSs are divided into many generation, network, and sales companies. This leads to the loss of administrative and economic integrity and manageability of EPSs, to the loss of economies of scale, and to many problems and negative consequences for electricity consumers.

Among the issues of electricity market organization the following are considered in the book:

- Difficulties and, often, deficiency of state regulation of the monopoly power companies. This was one of the arguments for deregulation of the power industry.

However, improvement of state regulation is a *real alternative* to deregulation, and a less risky one.

- Principal differences between electricity markets and markets in other industries, which are caused, as was already mentioned, by specific properties of an EPS (and electricity as a commodity).
- The choice of the model for electricity market organization with account taken of the features of certain countries (there is no "best" model acceptable for all the countries).
- The problems arising from deregulation of electricity markets (liberation of prices). There are many, but the main ones are the following:
  - *Increase in the wholesale electricity prices* from the level of average costs throughout the EPS (under price regulation) to the level of costs of the least efficient (marginal) plant. This leads to additional expenses for consumers and extra profits (so-called producer's surplus) for power generation companies (PGCs).
  - *Difficulties in financing* the construction of new power plants, including the "price barrier" to new power producers, which may cause capacity shortage and greater wholesale price increase. This will place a further burden on electricity consumers, whereas producers will start to get a monopoly profit.

One can also indicate a *general decrease* in power supply *reliability* with increasing probability of blackouts that have occurred lately in the USA, Canada, and Western Europe (and Russia). However, we will touch upon this only briefly.

The flaws and negative consequences of the *competitive* electricity market have been pointed out by many of its opponents. Over the past years they have increasingly manifested themselves in the countries which have organized such markets. Their reasons have to be sought for in the fallacy, groundlessness, or insufficient development of the concepts of competitive electricity markets. The main (apparently, not all) initial principles underlying the concept of competitive markets are presented in the box below. The author regards almost all of them as declarative (groundless) or erroneous.

The Initial Principles (Arguments, Postulates) of the Concepts of Competitive Electricity Markets

1. Conditions for *perfect competition* can be created in the wholesale and retail electricity markets.
2. Modern power systems *have lost* economies of scale. Therefore, vertically integrated power companies have *ceased to be* the natural monopolies.
3. State regulation *cannot become* efficient in terms of incentives for decreasing costs of electricity producers.
4. Competition in the wholesale market will result in *decrease* in wholesale electricity prices.
5. *Spot* electricity markets can be organized with real-time trade.

6. Market *itself* (without regulation) will ensure the adequate expansion of generation capacities in EPSs.
7. Creation of retail electricity markets is important to ensure "the right of consumers to choose a supplier."

For example, the author has not found in publications a profound enough analysis or substantiation of such points as:

- *To what extent the requirements or conditions for perfect competition in electricity markets are met.* In the statement that the competition reaps positive results, they always mean perfect competition. If there are no conditions for perfect competition in the industry, then the free (nonregulated) market will turn into an imperfect market—monopoly, oligopoly, monopsony, etc. Even a rather substantial book [23] in which the conditions for perfect competition are formulated has no analysis of their fulfillment in the electric power industry and almost the entire book was written on the assumption that the competition is perfect.
- *What factors create economies of scale in EPSs and whether these factors can cease to act (disappear).* There are references to combined-cycle plants which at small capacities can have better specific economic indices than large nuclear or conventional power plants on fossil fuel. However, this is a too narrow interpretation of the economies of scale for EPSs. It relates only to the economic efficiency of increasing the capacity of power plants and their units. In fact, however, this effect is integral and covers all the spheres of electricity production, transportation, and distribution in their interaction. The advent of combined-cycle plants, vice versa, enhances the efficiency of EPSs as a whole, with an increase in its total capacity or territorial expansion.
- *What kind of short-run and long-run* (in the microeconomic sense) *cost curves do power plants and PGCs have when they participate in the competitive market.* Supposedly, the shape of the cost curves of electricity generation is similar to that of "typical" firms in the other industries. However, it differs essentially in the electric power industry, and this explains the "troubles" with spot electricity markets and difficulties with investments in new power plants.
- *How do the efficiency and financing mechanisms for intersystem electric ties (including interstate ones) change with transition to a competitive market.* It is noted everywhere that network construction decreased sharply after deregulation; however, the reasons are rarely analyzed. Though these can also be explained by the distinctions of competitive market in the electric power industry.
- *What at all is the effect expected from transition to a competitive market,* i.e., whether the competition effect can exceed the costs for organization and operation of the competitive market and its potential negative consequences. Here it is necessary, of course, to differentiate between the effects for electricity producers and those for electricity consumers whose interests concerning price changes are totally different. The effect for consumers can only mean a decrease in electricity

price. Probably somewhere such a comparison, or a feasibility study, was made, but not in Russia, that is for sure. The comparison will be unfavorable, in particular, for the competitive *retail* electricity markets.

Thus, we can suppose that the original concepts of competitive markets in the electric power industry (which have been then repeatedly revised) were worked out without comprehensive analysis and consideration for properties and distinctive features of EPSs. Later when the disadvantages of competitive markets became noticeable, the main points of these concepts were taken as undeniable truth and were not thoroughly analyzed and revised. The belief was maintained that it is only necessary to properly design the market and it will operate well. However, in his research the author arrived at the conclusion that a normal competitive market in the electric power industry *is impossible by its very nature* and the efforts to organize and improve it will inevitably fail.

The information in the second box represents the main points showing the competitive electricity market failures which are explained in the book.

Main Points (Statements) Substantiated in the Book

1. In the electric power industry almost all the conditions (requirements) for perfect competition are not met and it is impossible to create such conditions.
2. EPSs as a whole are objectively characterized by the economies of scale which are constantly maintained unless they are forcedly split.
3. Characteristics of average (specific) short-run and long-run costs of electricity generation differ in essence from the same characteristics of "typical" firms considered in microeconomics.
4. In the electric power industry it is impossible in principle to organize spot markets ("trade on the spot") similar to the markets in other industries that could give price signals on production volumes and market expansion (or shrinkage).
5. With transition from the regulated to the competitive wholesale electricity market, the financing mechanisms for construction of new power plants essentially change. This creates obstacles to investments and a threat of deficit.
6. In the competitive wholesale market the equilibrium prices are formed at the level of costs of the least efficient (most expensive) power plant and these costs are always higher than the weighted average generation costs throughout the entire EPS.
7. Potential effect of competition in retail electricity markets is lower than the costs of their organization and operation.
8. In the competitive market electricity export ceases to be mutually effective. It is not beneficial for the consumers of the exporting country and the producers of the importing country.

9. Competitive (unregulated) wholesale market in the electric power industry is profitable only for electricity producers. The consumers, however, suffer due to increase in prices, deficits, and blackouts.

In order to show the imperfection of electricity markets and their distinctions from other markets, Chap. 2 discusses the EPS properties and specific features that are important for the organization of a market. Chapter 3 describes the main notions and types of markets that are applied and studied in the theory of microeconomics and their interpretation in the electric power industry. These two chapters can be considered an introduction for the rest of the book.

Chapter 4 presents the main models of electricity market organization, their comparison according to several criteria, and the flaws of the competitive electricity market based on the comparison. The expected competition effects and their possible involvement in electricity price regulation are analyzed as well.

Chapter 5 is devoted to the analysis of costs of electric power plants and PGCs as well as to price setting in the wholesale electricity markets in the short run. It also shows the failure of spot electricity markets and principal distinctions between the curves representing average costs of power plants and the curves demonstrating the costs of "typical" firms considered in microeconomics. The increase in wholesale prices while shifting to a competitive market and the formation of "producer's surplus" are illustrated.

Chapter 6 focuses on the problems of EPS generation capacity expansion (in the long run) under different market models. In the competitive market similar problems occur in financing the intersystem electric ties, which are discussed in the last section of the chapter.

Chapter 7 reviews the latest experience of power industry restructuring in different countries. The review discusses the countries with different market types. An emphasis is placed on the countries in which the competitive market has led to severe consequences.

The last chapter addresses the reforms of the electric power industry of Russia and the forecast of its development. The restructuring process starting in 1991 and the expected consequences of the transition to a competitive market in the nearest decade are analyzed. The recommendations are made to adjust the concepts of electricity restructuring in Russia.

All sections of the monograph end with a short summary presented in numbered boxes. A reader who wants to read the book "at a glance" can read these summaries and the conclusion.

Chapter 2
Electric Power Systems, Their Properties, and Specific Features

The electric power industry and electricity as its product are characterized by essential features and distinctions from other industries and commodities. Electric power systems (EPSs) underlie the electric power industry and determine its properties and the specific features of the electricity market. They are, in fact, the main subject of the studies described in the book and the focus of this chapter.

First, a general notion of EPSs (Sect. 2.1) and system effects (Sect. 2.2) that can be lost due to poor market organization is given. Special attention is paid to EPS properties that condition the specific features of the electricity market (Sect. 2.3). Section 2.4 presents briefly certain differences between EPS expansion and operation in the planned economy (in the USSR) and in the market economy.

2.1 General Definitions and Classification of EPSs

EPSs are rather diverse in terms of territory, degree of centralization in their management, composition of power plants, types of transmission lines, etc. In many countries and regions of the world there are also power interconnections that include two or more EPSs, interstate ones in particular, with their respective inter system ties.

The most general definition of EPS can be as follows [24]: An EPS is an integration of interconnected power plants, transmission lines, and consumer substations that are combined by the common process of electricity generation, conversion, transportation, distribution, and consumption. This definition is applicable to an EPS of any territorial level and any composition of power plants and transmission lines.

Let us just note that electricity produced by each EPS is a standardized product with a normalized frequency of alternating current, voltage, and some other indices. There cannot be several "types" of electricity on the territory of one EPS, i.e., it is not interchangeable for a consumer. A similar situation arises in water, gas, and heat supply systems.

L. S. Belyaev, *Electricity Market Reforms*,
DOI 10.1007/978-1-4419-5612-5_2, © Springer Science+Business Media, LLC 2011

An EPS to be considered as an individual ("independent") system should meet certain requirements (conditions):

- Maintaining a balance between electricity production and consumption with consideration of the required generation capacity reserves. Thus, the EPS should be *self-sufficient*, capable of operating separately from other EPSs.
- Unity of dispatching control (hierarchical, if needed) of the operation processes in normal and emergency conditions.
- Centralized forecasting, designing, and planning of power plants and EPS network development.

The first two conditions apparently do not need any special explanation. As for the united EPSs, they should be satisfied for the interconnection as a whole. In the course of development, some EPSs in the interconnection may become deficient or surplus.

The last requirement is of great importance for the subject of the book and needs to be commented upon. EPSs, as any other technical systems, should be designed and developed as an organic unity. Otherwise, they will not be able to perform their functions properly (in an optimal way). In the USSR, in the second half of the twentieth century, a special kind of economic and engineering activity emerged: "the electric power systems design" [25–29]. This activity favored optimal development and reliable operation of the Unified Power System (UPS) of the country.

In countries with a market economy, EPSs were also developed mainly in a centralized way within the regulated monopoly companies. After restructuring the power industry and transition to a competitive market (in some countries), expansion of generation capacities of EPSs was initially left to "an invisible hand of a market." However, experience and later studies have shown (see, for example, [19, 30]) that this "hand" does not always ensure adequate commissioning of power plants. Therefore, special measures were elaborated to contribute to deficit-free development of EPSs. These measures, in one form or another, provide for centralized forecasting of electricity consumption growth and "forced" (nonmarket) generation capacity expansion. As will be shown in Chap. 6, on the whole, transition to the competitive market complicates the fulfillment of the third condition, inevitably influencing further development of EPSs in the countries that have moved to the competitive electricity market (see Chap. 7).

EPSs and power interconnections were and are formed and developed gradually with transition to higher integration levels. These processes in different countries and regions of the world are highly diverse and depend on specific economic, geographical, and political conditions. On the whole, they are, however, objective and are caused by the effects of creating and interconnecting EPSs that are considered in the next section. Generally, EPSs can be classified as follows:

1. *Individual EPS*, which has been dealt with above.
2. *Interconnected EPS* (IPS) in one country. In the USSR, for example, there were 11 IPSs (Russia now has 7). The USA and China have several IPSs each.
3. *National or Unified EPS of the country* (NPS or UPS). These systems or interconnections were formed in the majority of European countries, in South Korea,

and many other small countries. The UPS of the country was in the USSR and is now in Russia. UPSs are being formed in China and India. In the USA, there is no UPS yet; however, almost all of their IPSs are connected to neighboring countries (Canada and Mexico).

4. *Interstate electric power interconnections* (ISEPI). These were formed in Europe, North and South America, in CIS countries, in the north of Africa, and in some other world regions. First of all, they aimed to trade in electricity with the neighboring countries, but in some of them other interconnection effects (system, capacity) have been achieved. ISEPIs are rather diverse in the number of countries involved, in terms of organization of their development management and operation control, the effects achieved, etc.

This book will address primarily interconnected and National (or Unified) EPSs in one country with a definite type (model) of electricity market organized. The international electricity markets created in ISEPI will be considered only at times to show how they are influenced by the types of market operating in the countries to be interconnected. In particular, the type of electricity market determines to a great extent the efficiency of intersystem electric ties (see Sect. 6.5).

Box 1 Classification of EPSs

1. Territorial levels of EPSs:
   a. *Individual ("independent") EPS,* which must be self-sufficient and balanced in electricity production and consumption
   b. *Interconnected EPS* (IPS) within one country
   c. *National or Unified EPS of the country* (NPS or UPS)
   d. *Interstate electric power interconnection* (ISEPI)
2. The book deals primarily with the EPS, IPS, and NPS (or UPS) located in one country with one or another type of electricity market organized.
3. In parallel with the unity of dispatching control, the *centralized* development forecasting and planning of the EPS, IPS, or NPS as an entity should be provided.
4. Electricity is a *standardized* product that is not interchangeable for a consumer on the territory covered by an EPS, IPS, or NPS (UPS).

2.2 Benefits of Creating and Interconnecting EPSs

It is widely known (see, for example, [16, 24–26]) that some objective reasons and factors have given rise first to creation of and increase in EPS capacity with extension of the territory served and then to expediency of their interconnection. On the whole, they impart a distinctive economic property to EPSs—*economies of scale*, i.e., an integral effect of decrease in costs of production, transportation, and distribution of electricity (and its price) with the growing EPS sizes. This property is seen in both individual EPSs and at their integration, encouraging creation of power interconnections of increasingly higher levels.

Let us consider at first the *factors contributing to the formation and expansion of EPSs*. Among them are the following:

- *Decrease in the required capacity reserves.* Increase in the total number of power units is known to decrease the probability of simultaneous emergencies of their specified share (percentage) (see, for example, [31]). As a result, the share of standby units to ensure the same reliability level of power supply is reduced with the growth of their total number. This concept is illustrated quantitatively in [24]. Dependence of the required emergency reserve on the total installed capacity of EPS proves to be nonlinear, namely, the reserve required increases to a lesser extent than does the total capacity of EPS. This objective regularity gave impetus to EPS formation, increase in EPS capacity, and territorial coverage as well as interconnection of EPSs.

  Here we note the following factors:

  - The considered effect is achieved by the increasing *number* of units regardless of their capacity, i.e., the "scale" in this case emerges in the growing number of units (blocks) of power plants, rather than in their capacity.
  - The effect is realized by the construction of *transmission lines* interconnecting power plants and consumer substations into the unified whole. Hence, this effect is typical of an EPS as a whole—in the interaction between the spheres of electricity generation and transportation (distribution).
  - With increase in the size (total capacity and area) of an EPS and preservation of its *integrity*, the effect will "fade away," i.e., decrease in the relative value, but continue increasing in the absolute one. This regularity can be violated by splitting the EPS into spheres and the spheres into several individual companies.

- *Improvement of specific economic indices of EPS facilities* with the enlargement of power plants and increase in transfer capabilities of transmission lines. This trend is well known. It showed up in the process of EPS dimensions growth, when it became possible (and economically sound) to construct power plants of higher capacity with larger units and higher voltage transmission lines. At present, the unit capacity of blocks of coal-fired steam turbine plants and nuclear power plants with thermal reactors has virtually reached its economic limit. Further increase in their capacity does not actually lead to decrease in their specific capital investments. However, it is still reasonable to construct such power plants with blocks of high (economically sound) unit capacity, if their commissioning is needed for the optimal EPS structure. Special place is occupied by hydropower plants (HPPs), whose capacity depends on concrete river conditions (water heads and flow rates), power plants on natural gas with combined-cycle installations (CCIs), whose sufficiently low specific investments can be achieved at low capacities of blocks, and also nuclear power plants (NPPs) with fast reactors, whose unit capacity has not yet reached an economic limit. Economic transfer capability of transmission lines, especially DC lines, can also increase.

Note that just this factor is often considered as economies of scale in the electric power industry. It is asserted, in particular (for example, in [32]), that with the appearance of CCIs the economies of scale have been lost. However, this is not so. First, this factor is one of many considered here. Second, the emergence of highly economic CCIs cannot lead to expediency of "destroying" EPSs or ceasing growth of their dimensions. CCIs, on the contrary, increase the variety of types of generation capacities and possibilities for creation of their more optimal structure, i.e., enhance an overall efficiency of electricity generation sphere, in particular at EPS expansion.

Construction of CCIs by independent power producers (IPPs) in regulated monopolies is a special case. The high efficiency of CCIs makes it possible for IPPs using them to successfully compete with monopoly companies. In this situation, it is obviously expedient to connect IPPs to the networks of the EPS that is owned by the monopoly company and conclude corresponding contracts on electricity supply. Such a condition is laid down by the Law in many countries (the USA, Japan, China, etc.). At the same time CCIs can be constructed by the monopoly companies themselves, which is practically the case.

- *Improvement of economic indices of EPS as a whole owing to the technological progress in any sphere of electricity production, transportation, or distribution.* The impact of technological progress is observed constantly and the EPS (as a system) "accumulates" the effects achieved in any sphere. Specific technological innovations are highly diverse. However, on the whole, they improve the EPS efficiency (reduce electricity prices and tariffs for final consumers) and contribute to the growth of their scales in both territory and capacity. Examples of the latest achievements in technological progress are the creation of the aforementioned highly efficient CCIs and the design of the FACTS (Flexible Alternating Current Transmission Systems), increasing transfer capability and controllability of AC transmission lines (see, for example, [33]).
  When an EPS is split into spheres and numerous independent companies, as is the case at transition to the competitive market, the effect of technological innovations can "remain" in the companies and not "apply" to consumers.
- *Optimization of structure, schemes, and operating conditions of EPSs*, whose possibility (and necessity) enhances economic efficiency of power supply to consumers, reduces costs in the system and electricity prices. Optimization implies selection of the most economically efficient power plants and transmission lines and the best modes of their usage. This factor, therefore, contributes to the formation of EPSs and assists their expansion (increase in EPS dimensions).
- *Decrease in the share of administrative expenses* with the growth of EPS scales, which is typical of vertically integrated companies that monitor the whole system.

Such a trend occurred everywhere in the last century. Nowadays, in the countries entering the competitive market, in which the single monopoly companies are split into sets of generating, network, and sales companies, these expenses have not fallen, but risen instead.

In general, as was already mentioned, the indicated factors create economies of scale, providing an incentive for the formation of EPSs, successive increase in their capacity, and territorial expansion. In the planned economy countries (including the USSR), this process proceeded under centralized management. In the market economy countries, in the first half of the twentieth century, it brought the *natural monopolies*[1] in the electric power industry into being that should be regulated by the State to prevent them from taking advantage of their monopoly position. Formation of the regulated natural monopolies was a structural transformation of the electric power industry of these countries in comparison with the earlier free market that existed there. The deregulation of the power industry taking place in some countries is a reverse transformation (return to the competitive, though institutionalized, market). The present book is devoted exclusively to the analysis of this transformation.

Now, we pass on to the *effects owing to the interconnection of EPSs* with the formation of IPSs and the NPS (or UPS). These effects are also well known and studied (see references indicated at the beginning of this section). Therefore, they will be commented on briefly. Part of the effects is due to the same factors that were mentioned above; however, there are specific factors as well.

The key effects achieved owing to the interconnection of EPSs are as follows [24]:

1. Power transfer from an EPS with cheaper electricity to an EPS with a more expensive one
2. Reduction in the required emergency and repair capacity reserves
3. Decrease in coincident maximums and leveling of the joint load curves of consumers
4. Possibility of construction of larger-scale power plants with larger units
5. Rationalization (coordination) of putting into operation large power plants in EPSs to be interconnected
6. Improved usage of power plants when interconnecting EPSs with a different structure of generation capacities
7. Environmental, social, and other effects

Decrease in the necessary emergency reserves (point 2) and the possibility to construct larger power plants (point 4) were also important in the creation of individual EPSs. The rest of the effects may be treated as specific that emerge when

[1] A natural monopoly is an industry in which the economies of scale is so high that the product can be manufactured by one firm at lower average total costs than if it was manufactured by several firms [34].

interconnecting EPSs. In concrete IPSs or NPSs, not all the enumerated effects, but only a combination of them or even only one key effect can naturally be found.

Each effect has to be estimated in monetary terms (in rubles, dollars, etc.) in one way or another, and if their sum exceeds the cost of an intersystem electric tie (ISET), it is advisable to interconnect EPSs. As a rule, the economic assessment of the effects, in particular the environmental and social effects, proves to be difficult enough. It requires special calculations on the basis of appropriate mathematical models [24].

Note that the specific features of realizing different effects are important for a further study of the electricity markets. These features are stipulated in particular by the fact that many effects owing to the interconnection of EPSs are expressed in generation capacity saving, and are achieved by the construction of intersystem transmission lines. Some market models propose separation of the spheres of electricity generation and transmission (and distribution) and creation of independent generation and network companies. In this case, thc network companies will bear the costs and the generating companies will take advantage of the effect. Such an inconsistency (in comparison with single vertically integrated companies) will complicate the substantiation of the ISET efficiency, and hence the interconnection of EPSs (see Sect. 6.5).

Transmission (export) of cheap electricity from one EPS to another will shift the construction of new power plants, and, as a result, the first EPS will become surplus and the second deficient. At the same time it may influence electricity prices: they can fall in the receiving EPS and, on the contrary, rise (electricity demand will increase) in the transmitting (exporting) one. In different models of electricity market organization these factors will show up in different ways. In the markets with regulated electricity prices, such an export may be mutually beneficial if the export price is set within the range of prices of EPSs to be interconnected. Then the consumer price can be reduced in the exporting system owing to the export earnings, and in the receiving system owing to cheaper electricity received. In competitive markets with free prices, electricity export will cause a loss to consumers of the transmitting system because of increase in electricity demand and prices (see Sect. 6.5).

The following two types of effects—decrease in the required reserves and coincident maximum load (in comparison with the sum of maximums for EPSs at their isolated operation)—directly lead to savings in generation capacities. They may be called "capacity" effects of interconnecting EPSs. These effects are very substantial for some countries. Their quantitative assessments for the UPS of the USSR are presented in Sect. 2.4. They are typical of the EPS as a whole at a joint consideration (efficiency assessment) of the electricity generation and transmission spheres, when construction of transmission lines decreases demand for generation capacities of EPSs to be interconnected and the total costs for EPS expansion.

The capacity effects of interconnecting EPSs are observed at any type of generation facilities and transmission lines. This fact is often underestimated when one speaks of the loss of the economies of scale in the power industry. The economies of scale implies not only economic feasibility of increasing power plant sizes and transfer capability of transmission lines. It is typical of EPS as a system, i.e., costs in the transmission sphere decrease costs in the electricity generation sphere. It cannot disappear and will constantly manifest itself with the increase in EPS scales, if it is not split into spheres and sets of companies.

The considered three types of effects also occur when interconnecting EPSs of different countries and creating ISEPIs. The intensive formation of interstate power interconnections in almost all world regions proves it [24]. Hence, the economies of scale is inherent in EPSs both at the national and interstate levels.

The rest of the effects will not be commented upon. As a rule, their realization depends to a lesser extent on the electricity market type. They are described in greater detail in the mentioned works, in particular in [24].

Box 2 Efficiency of EPS Creation and Interconnection

1. EPSs are characterized by the economies of scale that implies decrease in costs of electricity production, transportation, and distribution and *its price* with the EPS expansion. It is specified by many factors and was the main economic incentive in the formation of up-to-date EPSs and their subsequent interconnection.
2. Owing to the economies of scale, *the regulated natural monopolies* were formed in the middle of the twentieth century in the electric power industries of countries with market economies. They ensured a fast growth of electricity consumption and the appropriate expansion of EPSs. There are no reasons to suppose that in the current EPSs the economies of scale have changed into the diseconomies and the electric power industry has ceased to possess properties of a natural monopoly.
3. Interconnection of EPSs yields rather diverse effects that can be estimated one way or another in monetary terms. If the total economic effect exceeds the costs of construction of an intersystem tie line, interconnection of EPSs is expedient. The interconnection effects have led to the extension of EPS areas over the whole country and to the formation of interstate interconnections.
4. Part of the effects (capacity effects) consists in decrease in the necessary generation when EPSs are interconnected. In this case the final effect is achieved due to interaction of electricity generation and transportation spheres, i.e., for IPS, NPS (UPS), or ISEPI as a whole. Compulsory separation of these spheres can cause problems in achievement of the capacity effects.
5. In general the effects of both EPS creation and interconnection are a reason for the economies of scale in the electric power industry. Undoubtedly, it is a favorable property that contributes to the reduction of the total production costs and prices of electricity. Its loss at power industry reform should be treated as a disadvantage.

2.3 Properties of EPSs

Properties of large energy systems (LESs), including EPSs, have been studied to a great extent (see, for example, [35–37]). Main attention was paid to the common properties characteristic of all or several types of LESs: hierarchical structure, uncertainty of initial information, reliability, dynamics, etc. These studies resulted in the elaboration of certain methodological approaches, principles, and methods to be applied for development and operation management of different types of LESs. In parallel, specific types of systems, including EPSs, undoubtedly have individual properties appropriate only to them. As a rule, some scope of common (for LESs) and individual properties proves to be important for one or another aspect of specific large system management, individual properties being often of crucial importance.

Sets of physicotechnical, economic, social, and environmental properties are surely typical of EPSs. In our discussion below, consideration is given to those influencing market organization in the power industry in one way or another. Based on the variety of possible market types (models), the display of these properties will be noted in different (and sometimes in all) market models.

Chapter 4 deals with models of electricity market organization in detail. Here, the most general idea about them seems to be expedient for further illustration of the impact of different properties of EPSs on them.

Figure 2.1 presents four major models of the electricity market [32, 38]:

1. *Regulated natural monopoly* (absence of competition), which was already mentioned above. In the electric power industry these are the so-called vertically integrated companies embracing all the spheres of electricity production, transportation, distribution, and sale. This market form has given rise to restructuring or reform discussed in the book. The following market models are characterized

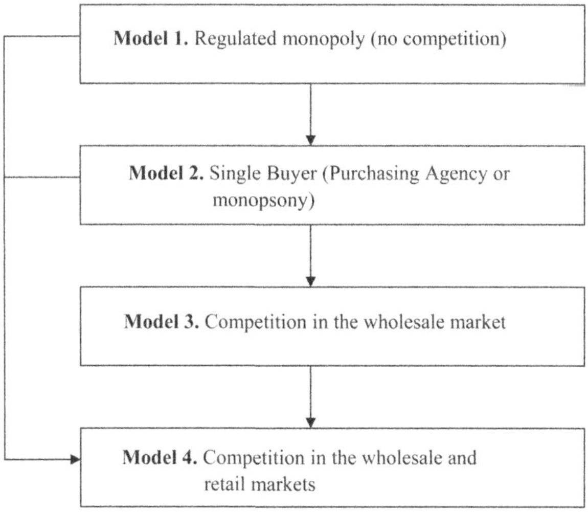

Fig. 2.1 Major models of electricity market organization

by successive separation and differentiation of the indicated spheres with formation of the corresponding generation, network, and sales companies.

2. *Single buyer* (Purchasing Agency, monopsony), when the generation sphere is divided into several separate (financially independent) power generation companies (PGCs) that start to compete with each other in electricity supply to the common Purchasing Agency. The other spheres remain vertically integrated in the agency and it is a monopolist with respect to consumers as before. Business of the Purchasing Agency, therefore, should be regulated by the State, including price quotation of electricity purchased from producers and sold to consumers.
3. *Competition in the wholesale market*, when the electricity transportation sphere is separated, the spheres of electricity distribution and sale are split into territories and the wholesale market is organized. This leads to creation of transportation-network company, territorial distribution-sales companies (DSCs), and specialized market structures. The wholesale market prices become free and the activity of DSCs and the retail prices are regulated as before.
4. *Competition in the wholesale and retail markets*, when the spheres of electricity distribution and sale are additionally divided with formation of regulated distribution companies (by territory) and sets of independent sales companies. Retail electricity markets are organized with competition between sales companies (buying electricity in the wholesale market) and consumers. The retail prices are no longer regulated.

We should underline that all the enumerated models are market models, as often only the last two models are called markets. The nonmarket electric power industry under a planned economy will be discussed in the next section. The first two models are markets with regulated prices—tariffs[2]—and we will call them, for short, regulated markets, while the third and fourth models will be markets with free prices or competitive markets. For brevity's sake, these models will sometimes be referred to by the numbers under which they have been listed above (Model 1, Model 2, etc.).

The arrows on the left in Fig. 2.1 show transition at restructuring from the regulated monopolies at the regional level and the single-buyer model at the federal level to Model 4. The transition is stipulated by the Law of the RF "About electric power industry" [39] (for more details see Chap. 8).

Now we will address directly the *properties of EPSs*, which determine specific features of the electricity market.

The following are the well-known properties and features of EPSs:

- A special role of electricity in the economy and society; damage caused by sudden interruption of electricity supply exceeds manifold the cost of undersupplied electricity, which requires special measures to support electricity supply reliability.

[2] To be more precise, the regulated prices will be called "tariffs" as opposed to market prices formed in the competitive markets.

- Inability to store (accumulate) electricity in rather large volumes.
- Necessity to balance electricity production and consumption at every moment.
- Inevitability of equipment failures, and hence the necessity of backup generation capacity and electric ties.

These properties undoubtedly influence and complicate market organization in the power industry to a varying extent in different market models. However, note some other features of EPSs that are also important in this context and are interrelated with the above properties in one way or another:

1. *Specialized electricity transport* (by wires). It excludes electricity delivery by general types of transport (railway, motor, water, air), which is possible for production of the majority of other branches and renders a local character to EPSs. New electricity producers and consumers can emerge only by connecting them to EPS networks. This property leads to:

- Territorial limitedness of the electricity market: only consumers and producers directly connected to the EPS through electric ties with a sufficient transfer capability are able to participate in the market. In particular, there is no world electricity market or world electricity prices.
- Participation of only existing (operating) power plants in the market.
- Existence of the technological (physical) barrier to the entry of new producers into the market; to this end new power plants should be constructed and connected to EPSs. Thereby, one of the principal conditions of *perfect* competition—*free entry of new firms into the industry and free exit of existing firms from it* [34]—is not observed in the power industry.

It should be noted that a physical barrier for new power producers (NPPs) is especially important. It plays a decisive role in electricity markets in the short run (in the microeconomic sense; see Chap. 3). NPPs simply cannot appear in the market, because a new power plant should be designed, constructed, and connected to the EPS, which requires several years. In the short-run electricity market the operating producers are protected from competition of NPPs and can raise prices. It is one of the basic reasons for electricity market imperfection and it cannot be eliminated (i.e., it is impossible to make the market perfect) by any organizational and methodological measures or rules.

2. *Daily, weekly, and seasonal load variations* that determine:

- The need to expand generation capacities in accordance with an annual load peak (taking into account reserves); in other periods of the year power plants will be underloaded and get lower revenues which may turn out to be insufficient to pay back investments.
- The economic expediency to have different power plants (basic, peak, and semipeak) with various economic indices (specific capital investments and production costs).
- The need to optimize the structure of generation capacities (by type of power plants) and operating conditions of power plants for different time periods of a year.

The presence of power plants of different types, in turn, leads to specific supply curves of producers and to formation of marginal prices and producers' surplus [40] for more efficient power plants in the competitive wholesale market (for more details, see Chaps. 3 and 5).

This feature of EPSs also caused the need for centralized dispatching control of normal and emergency operation of the power system (which is foreseen in all market models) and also engendered the next property (or even paradox) in the electric power industry which is observed in no other industry.

3. *The need for optimization of the power system operation with regard to instantaneous (hourly) variable costs of power plants, while their total costs (and economic efficiency) are determined by integral operation results for the whole year with an account taken of fixed costs.* Load variations during a year cause changes in operating powers (load) of power plants, which should be optimized according to the criterion of the least hourly, daily, weekly, or seasonal variable (fuel) costs throughout the entire power system. While carrying out optimization we have to use hourly characteristics of power plants, which represent only variable costs.

Meanwhile, the real electricity value (and its price) is determined by the average total costs, including fixed costs of power plants as well. In the electric power industry the average total costs can be determined only for the whole year. They will depend on an annual output of a power plant, its operation during a year (which determines annual variable costs), and annual fixed costs. This difference between hourly and annual costs influences essentially the organization of electricity markets and the process of price setting. In particular, the spot electricity markets organized in real time (with hourly or half-hourly intervals) are not real short-run markets considered in microeconomics, and their prices do not reflect the real value of electricity, which makes the spot markets inappropriate (Sects. 5.1 and 5.2). The real short-run electricity markets can only be the markets that cover the period of one or more years and are implemented through respective contracts.

4. *Great capital intensity, long periods of construction, and service of power plants and some transmission lines, which result in:*

- The impossibility of quickly eliminating shortage if it occurs for some reason. It will take several years to design and construct new power plants. Moreover, if power plants are constructed by private investors (Models 3 and 4), nearly 10 years more will be necessary to pay back the investments. Hence, private investors should know the power system expansion conditions, including the prices in the wholesale market, 15–20 years in advance. These conditions are rather uncertain, which create a large risk for investors and make construction of new power plants and elimination of shortage even more complicated.
- The need for prior planning and subsequent financing for the expansion of generation capacities in power systems to avoid shortage in the electricity market.
- Power plant service life (30–40 years) exceeding "reasonable" payback periods (10–15 years), which will make private investors construct power plants (Models 2–4).

This feature of EPSs manifests itself to a greater extent under competitive markets (Models 3 and 4) when the criteria, incentives, and financing mechanism for construction of new power plants change dramatically as compared to the regulated monopoly and single-buyer market. These changes create problems of investing in the expansion of generation capacities, which are considered in [19] and in Chap. 6.

Moreover, the competitive market concepts (including those in Russia) usually envisage no centralized planning of the generation capacity expansion. The generation capacities are supposed to expand on the basis of "market signals." However, as was pointed out in Sect. 2.1, the experience of the countries that introduced the competitive electricity market and recent research have shown that the market does not generate these signals timely and special "non market" measures are required to prevent power shortage.

5. *High level of mechanization, automation, and even robotization (at nuclear power plants) of electricity production, transportation, and distribution.* Normally, power plants and substations have only administrative, duty, and maintenance personnel. The number of personnel practically does not depend on the amount of actually generated and transmitted power. All process lines and units at power plants are designed on the basis of their maximum (installed) capacity.

This feature of EPSs along with the said huge capital intensity of power plants leads to a high share of fixed costs in the total electricity production costs. At the same time, there are practically no variable costs at HPPs, and those at nuclear and thermal power plants are made up of fuel costs only. The characteristics (curves) of average costs of power plants, therefore, differ principally from the cost curves of "typical" firms considered in the theory of microeconomics (see Chap. 5). This makes the short-run competitive wholesale electricity market "nonstandard," i.e., different from the markets in other industries. In particular, power plants (or power generation companies) will have to enter the market with their supply bids reflecting the total costs rather than the marginal ones.

6. *Interdependence of electricity production processes of different power plants in the power system.* All power plants operate to cover the total EPS load which changes daily and seasonally. Their operating conditions are optimized centrally, depending on the mix of generation capacities in the EPS.

This feature of the power system brings essential peculiarities in the electricity market:

- Power producers (sellers) do not enter the market with already finished products with known volumes and prices. Electricity is produced *jointly and simultaneously* by all producers. Volumes and costs of each producer will depend on centrally assigned operating conditions for different hours, days, and seasons. The most economically important *annual* volumes and costs of each producer will be determined only at the end of the year by integral results.
- Thus, the *uncertainty* exists in the characteristics of short-run costs of power producers. This uncertainty is not observed in the industries where firms (companies) produce commodities *independently* of one another. The uncertainty of power plant costs makes the electricity market very special. In the regulated

markets (Models 1 and 2), this creates difficulties in establishing tariffs by the regulatory bodies. The regulation should envisage adjustment of tariffs in the event that the actual output of power plants deviates considerably from the planned one (this is particularly necessary for HPPs, whose output depends on random inflow of water). In the competitive markets (Models 3 and 4), the situation is even more complicated—the electricity producer in the market does not know exactly how much electricity he will produce throughout a year and what total costs he will bear. Naturally, he will overestimate the prices both in the spot market (if it exists) and in the long-term contracts with buyers.

7. *Facility-by-facility expansion of power systems.* The market in any power system expands through the construction of individual new power plants and transmission lines. This property reveals itself differently in different models of electricity market organization.

New power plants can be funded and constructed by:

- Vertically integrated companies (VICs) (Model 1)
- Power generation companies (PGCs) (Models 2–4)
- New independent power producers (IPPs) (Models 1–4)

As is shown in Chap. 6, financing mechanisms for the construction of power plants can vary. The primary distinction is that under regulated markets (Models 1 and 2) the investments in new power plants are paid back at the expense of the total electricity output generated by VICs (or in EPSs), whereas under the competitive wholesale market (Models 3 and 4) the investments in some power plants should be paid back at the expense of the electricity generated by only that one power plant.

Under the competitive market, each new power plant constructed by a private investor, along with operation costs, will have its own investment components required to pay back the investments. Therefore, the price to be offered by the new electricity producer in the wholesale market will be higher than the price offered by the operating power plant of the same type. This creates an economic (price) entry barrier for new producers in addition to the physical barrier mentioned above, which makes the electricity market imperfect in the long run as well (Sect. 6.4).

Additionally, the facility-by-facility expansion of generation capacities in EPSs influences the shape and sense of the long-run cost curves of the electricity generation sphere. Under competitive markets the short-run costs of new power plants should be considered as long-run production costs of IPPs and PGCs (Sect. 6.3).

Moreover, the transition to the competitive wholesale market changes the mechanism of financing the intersystem and interstate electric ties, which makes it difficult to substantiate their efficiency (Sect. 6.5).

8. *Economies of scale.* This was already considered in the previous section. To the greatest extent, this effect is realized in the regulated monopoly (Model 1). In other models, it subsequently decreases (Model 2) or is even lost completely (Models 3 and 4) due to the splitting of one company into several separate companies. It should be emphasized once again that this effect is typical of the entire EPS (as

a system) and not only of power plants in the electricity production sphere as it is sometimes interpreted (for example, in [32]).

The overall analysis of power system properties shows, on the one hand, the principal distinctions of the electricity market from the markets in the other industries and, on the other hand, its obvious imperfection.

The main distinctions are:

- Territorial limitedness of the electricity market (within the territory covered by the networks of a specific EPS).
- The need for dispatching control of normal and emergency conditions of the power system.
- The need for centralized design and advance planning of the power system expansion with account taken of the required capacity reserves.
- Impossible to organize "normal" electricity spot markets (for more details, see Sects. 5.1 and 5.2).
- Nontypical and uncertain costs in the generation sphere of EPSs, which makes the competitive (unregulated) wholesale electricity market "nonstandard" in the light of the theory of microeconomics (for more detail, see Sects. 5.3 and 5.4).
- Obvious uniqueness of intersystem electric ties that connect different territorial electricity markets (for more details, see Sect. 6.5).

The electricity market imperfection is first of all conditioned by the technological (physical) barrier to new producers in the short run and by the price (economic) barrier to them in the long run (Sect. 6.4). Whether or not the other conditions (requirements) of perfect competition are met is analyzed in Sect. 3.2. The imperfection of the electricity market reveals itself under any models of its organization. In Models 1 and 2, its monopolistic character is obvious and this leads to the necessity to regulate electricity prices (tariffs). In Models 3 and 4, as will be shown in Chaps. 3 and 6, electricity producers on the one hand may form an oligopoly[3] and on the other hand maintain "market power," thus having the chance to create shortage and raise electricity prices through cessation or delay in construction of new power plants. This is also facilitated by the economic barrier mentioned above.

It should be noted that the electric power industry differs from other infrastructural industries, such as transport or telecommunications, in the production of commodities. It is the sphere of electricity generation that creates many of the foregoing EPS distinctions and makes the electricity market imperfect. This, in particular, relates to a nontypical character and uncertainty of costs in the sphere of EPS generation, to the impossibility of organizing electricity spot market, and to the existence of physical and price barriers to entry of new producers into the market. It is important to indicate this distinction, since in some countries (for example in the USA) one of the arguments for deregulation of the electric power industry was successful reforms in the air transport and telecommunications. This distinction of the power industry is analyzed in [3].

[3] Oligopoly is a market form in which a small number of sellers dominate in selling a certain product and the entry of new sellers into the market is either complicated or impossible [41].

Box 3 EPS Properties and Their Impact on the Electricity Market

1. Certain properties and specific features inherent in power systems cause *imperfection* of the electricity market and its distinctions from markets in other industries. In addition to the well-known EPS properties, it seems advisable to underline the following ones:

   (a) *Economies of scale* considered in Sect. 2.2 that can be lost due to poor market organization.
   (b) *Specialized electricity transportation* that leads to territorial restriction of the market and creates a technological (physical) entry barrier for new producers in the short run.
   (c) *Variability of daily, weekly, and seasonal loads of consumers.* In combination with the other properties analyzed in this section, this one determines, in particular, the:

   - Uncertainty of short-run costs of electricity producers.
   - *Impossibility* of organizing "normal" spot electricity markets (for more details, see Sects. 5.1 and 5.2).

   (d) *High level of mechanization, automation, and even robotization of processes.* As a result, the characteristics (curves) of the average costs of power plants differ essentially from the cost curves of the "typical" firms studied in the theory of microeconomics (See Sects. 3.1 and 5.3). This fact renders a *short-run wholesale* electricity market to be "nonstandard."
   (e) *High capital intensity, long period of construction and service of power plants, facility-by-facility expansion of EPS.* These properties lead to:

   - Impossibility of fast elimination of *shortage.*
   - Necessity of advance *expansion planning* of EPS generation capacities to prevent shortage.
   - *A price barrier* for new electricity producers in the competitive market in the long run (for more details, see Sect. 6.4).

   It should also be noted that the electricity market type (model of organization) radically influences the *mechanisms of financing* construction of power plants (Sects. 6.1 and 6.2).

2. The key *distinctions* of the electricity market from markets of other goods that are worth considering are:

   (a) Its territorial restriction and the absence of the world electricity market and the world prices.
   (b) The necessity of dispatching control over operating conditions and centralized EPS expansion planning.
   (c) "Nontypical" character and uncertainty of costs in the EPS generation sphere.

   (d) Economic incompatibility of organizing spot markets of electricity.
   (e) Unique nature (in the microeconomic sense) of intersystem electric ties connecting different territorial markets (see Sect. 6.5).
3. *Imperfection* of the electricity market (in terms of microeconomics) creates, first of all, a technological entry barrier for new producers, and transition to the competitive electricity market leads to creation of a price barrier for them. The monopolistic position of electricity producers in the market is retained in all models of market organization, except for the single-buyer model (for more details, see Chaps. 3, 4, and 6).
4. *Commodity production* distinguishes the electric power industry from "purely" infrastructural branches (such as transport or telecommunications). The EPS generation sphere causes the aforementioned distinctions and imperfection of the electricity market.

2.4 Electric Power Industry in Planned and Market Economies

This section will address the distinctive features of the development and operation of the electric power industry in planned and market economies, the issues of state and private ownership, including corporatization of power companies, and some aspects of Russia's and China's power industry transition from the planned to the market economy.

Under planned economy, there is naturally no market in the power industry or at best there can only be some of its elements. In the USSR, in particular, the development and operation of the power industry were planned centrally along with other sectors of the economy. The plans were funded from the national budget and implemented by a hierarchical system of state organizations and enterprises, with the Ministry of Energy of the USSR at the upper level. In China, a similar centralized management of the power industry continued till 1985 or even up to 1997 when the Ministry of Power Industry was abolished [42].

Very often, the situation in the power industry under the planned economy is compared to the regulated monopoly under market conditions. There are, undoubtedly, some common points, but there are also considerable differences. These concern, first of all:

- The mechanisms of planning and regulation
- The establishment of electricity prices
- The types of ownership (state, municipal, private)

The differences between planning and regulation reveal themselves, to a greater extent, in the development of the power industry. In the USSR, for example, planning the development of industries was an integrated process that covered various interrelated time periods: perspective planning for 10–15 years, 5 years, and the 1-year plan. In the electric power industry this process was based on the designing of the power systems, i.e., the elaborated development strategies of UPSs and IPSs, feasibility studies on the most crucial issues, and the like. The plans for the development of the power industry suggested providing the national economy with electricity (and heat) on the one hand, and the power industry itself with the required resources including the financial ones on the other hand. Owing to the directive character of plans, particularly those annual ones, the uncertainty of forecasting electricity demand, commissioning of power plants, transmission lines, etc., was minimum. Development of the electric power industry (UPS) was planned for the whole country with regard to the development of individual regions (IPS). Here, due to the general trend toward minimization of expenditures, the effects of EPS interconnections were realized to the maximum extent, the minimum required level of capacity reserves was maintained, and so on.

Regulation of EPS expansion within the natural monopolies differs from planning in several aspects. First of all, the proposals for expansion of power systems on their territories are made up, as a rule, by the monopoly companies themselves. The regulatory bodies then coordinate (or correct) these proposals. This implies that:

- The work on analysis and optimization of power system expansion is done twice—by the monopoly company and by the regulatory body; the latter should be staffed with highly qualified personnel.
- The monopoly companies, particularly the private ones, are interested in the maximum expansion of their power systems, both to guarantee reliable power supply and to increase their capital, which is materialized in new energy facilities.
- The regulatory bodies are subject to constant pressure from monopolistic companies and, therefore, should be protected from corruption; simultaneously, being also responsible for power supply to consumers, they are inclined to allow surplus generation capacities rather than their deficit.

Secondly, expansion of power systems is planned and carried out by the monopoly companies, yet only within the territories they serve. If there are several such companies in the country, as for example in the USA, Canada, and Japan, it becomes difficult or even impossible to realize the effect of interconnection of individual power systems and create the most efficient Unified (or National) Power System of the country. This also leads to different electricity prices (tariffs) throughout the country, which will be discussed in more detail below.

Thirdly, it should be noted that power system expansion planning in the market economy is related to a much greater uncertainty than under the planned economy.

As to the control of power system operation, in particular dispatching control, it differs in planned and market economies to a lesser extent, and so we are not going to dwell on this. It can only be mentioned that the planned economy makes it possible to organize centralized dispatching control of the UPS (or NPS) of the whole country, with an appropriate increase in economic efficiency and reliability of the power supply. However, this is not feasible, generally, in the case of market economy countries with several monopoly companies in the power industry. In the USSR, the level reached in the development of methodology and facilities for hierarchical dispatching control of the UPS, including emergency control systems, was high indeed [43–48].

The principles and methods of establishing electricity prices (tariffs) in the planned economy and regulated monopolies differ greatly. In the USSR the prices were established on a centralized basis and were common throughout the country.[4] Moreover, taking into account the centralized planning and funding, the power industry could both be subsidized and a profitable industry. The authors in [49] and [50] show that in 1960–1965 the power industry did not recoup its expenses completely and then up to 1990 was profitable despite large capital investments allocated for its development. Here the electricity tariffs for industry and for the population (higher) were quite acceptable.

In the regulated monopolies the tariffs are established by the regulatory bodies individually for each company. The tariffs include all the required costs of the company, investment components, ensuring power system expansion, and normal profit. Sometimes, at the request of state or municipal authorities, the tariffs include some "indirect" costs, or, vice versa, the subsidies are allocated, for example, to implement renewable energy sources. The mix of generation capacities in the power system, fuel types and cost, as well as other conditions in each monopoly power company can naturally be different, which results in different electricity tariffs throughout the county if there are several such companies. The techniques of establishing tariffs, especially in terms of investments and profit, are country-specific. On the whole, the process of tariff regulation is rather complicated and affects enormously both tariff value and the financial state of regulated power companies. The difficulties and flaws of state regulation of monopoly power companies were, as already mentioned, one of the arguments in favor of transition to the competitive market.

In Russia, at the beginning of the 1990s when the electric power industry was split into joint stock companies, the state regulation of newly created power companies (RAO "EES Rossii," AO-Energos, AO-Electrostantsiyas) had to be organized anew and urgently. Naturally, there were many drawbacks aggravated by a general economic crisis in the country. Later, the system of regulation improved slowly, but surely it is still far from completion.

As to the property in the power industry, under planned economy (in the USSR and up to 1985, in China) it was 100% state-owned. In the regulated monopolies it can be state-owned (or municipal) (France, Norway, and some other countries) or privately owned (Japan, England, many states of the USA, etc.). In any case, the activity of monopoly power companies is regulated by the State. Starting in the 1980s,

---

[4] There can be different viewpoints on the expediency of equal electricity prices throughout the country. However, the efforts of Federal bodies in the USA and the EU indicate to their endeavor to equalize the prices among the countries (states).

private independent producers that sold electricity to monopoly companies at regulated tariffs emerged almost in all the countries (England, the USA, China, etc.).

In some countries, power industry restructuring was accompanied by the privatization of the generation companies, and in others they remained under the state ownership. Electric networks were not privatized, as a rule.

The process of separation of economically independent generation (and other) companies from monopoly vertically integrated power companies, with them remaining in the state ownership, is called "corporatization" [2]. Such corporative (state) companies act in the market similarly to the private companies, i.e., they also seek to gain maximum profit and compete with one another and with the private companies. Corporatization can be performed at different extents of power industry restructuring. For example, China and South Korea have corporate generation companies at the single-buyer market (Model 2), while Norway has them at the competitive market (Models 3 and 4).

There is an opinion that private companies are more efficient than state-owned ones. At the same time, the latter have a number of advantages over private companies:

- Their costs do not include "normal" profit to pay dividends to shareholders (owners) of the company; hence, they can enter the market with lower prices than similar private companies.
- Profit gained is disposed under the direct control of state bodies, i.e., in the interests of the State.
- A company's administration wage is also established by the State; it certainly depends on the results of the company activity (including the profit obtained) but cannot be established by the company itself. Simultaneously, the salaries of company employees are regulated, though in private companies the salaries of employees can be understated.

The above advantages of state companies may turn out to be more significant than the possible advantages of private companies. For example, Noble Prize Winner Professor Joseph Stiglitz shows [51] that the managers of private companies cease to pursue the interests of shareholders (lose touch with the owners). They have more detailed information concerning the market and the situation in the companies managed by them, and they use it for their selfish purposes.

In the conditions of state ownership under centralized planning and financing, the UPS of the USSR was the most integrated and efficient power interconnection [52]. It realized to a greater extent the effects of power system interconnection considered in Sect. 2.2, that is:

- The effect of peak load diversity among interconnected power systems (situated in six time zones) resulted in the decrease in demand for generation capacity in December 1991 by 6 GW [52]. At the same time in each IPS this effect reached 2.33 GW in the Central IPS, 0.91 GW in the Northwestern IPS, 1.01 GW in the Southern IPS, 0.43 GW in the IPS of Middle Volga, 0.59 GW in the IPS of Northern Caucasus, 0.37 GW in the IPS of Transcaucasus, 1.07 GW in the IPS of Ural, 0.32 GW in the IPS of Northern Kazakhstan, and 0.92 GW in the IPS of Siberia, i.e., in total another 8 GW.

- The saving of emergency and maintenance reserve in the UPS of the USSR is estimated at 3–4% [25].
- The effect of power plant utilization improvement (optimization of operating conditions) is estimated at 10–12 million tce saved yearly [52].

According to the gross estimation of the USSR UPS efficiency [25], the expansion and operation cost saving surpasses the costs of the UPS backbone network development by a factor of 1.5–2.5 (owing to the saving in the generation sphere).

On the whole, the UPS of the USSR, after its formation in the 1960s, ensured reliable and economical power supply to the country at the minimum required capacity reserves, optimal structure, and operating conditions of generation capacities.

After the USSR disintegration, the UPS was partitioned and many power systems of the CIS turned out to be deficient, and in the UPS of Russia the above effects considerably decreased. The reform of the power industry in Russia, which is currently underway, will cause further losses of the effects which were inherent in the UPS of the USSR (Chap. 8).

After the economic reforms launched by Den Siao Pin, China's power industry underwent dramatic changes. Introduction of market relations (and private ownership) at maintenance of the general centralized planning (and state property), i.e., transition to the mixed economy, resulted in an extremely good effect for China. The reform of the power industry in China will be described in detail in Chap. 7. Here, we will dwell only on the following circumstances.

Starting in 1985, China created very favorable conditions for the attraction of private investments into the sphere of generation. This was caused by the lack of own (state) funds to meet a speedy increase in power consumption and eliminate power shortage. As is shown in [19] and will be considered in Chap. 6, high rates of expansion of generation capacities make it profitable to involve credits for financing. In fact, the long-term contracts that were concluded with private investors and guaranteed payback of investments at a rather high interest rate represented a kind of these credits. This measure was economically feasible for China. It resulted in the emergence of many independent power producers (private owners) along with preservation and expansion of power plants that belonged to the state and municipal authorities.

Planning of power industry development had been performed by the Ministry of Electric Power Industry before 1998 and then by the State Commission for Development and Planning and was gradually replaced by state regulation. In 2002, power generation was separated from the transmission sphere (with formation of the respective state companies), and the Chinese Commission for Regulation of Power Industry was established. This commission controls development plans and establishes tariffs for the state companies, approves long-term contracts with independent power producers, etc.

Currently, China's power industry has been transformed into the regulated single-buyer market. Two companies have been founded: China's State Power Grid

Corporation, which covers the territory of the interconnected power system of Northern, Northeastern, Northwestern, Eastern, and Central China, and the South China Grid Company on the territory of southern and southwestern provinces of China. These companies, along with development planning and operation of electric network, perform dispatching control of power systems, plan expansion of generation capacities, hold tenders, and conclude long-term contracts for the construction of new power plants and power supply from them.

Thus, China has made a "gradual" conversion of the electric power industry under planned economy first to the regulated monopoly and then to the regulated single-buyer market. At the same time, there are both state (and municipal) and private PGCs that compete with one another. The system of state regulation provides, on the one hand, maximum possible construction of new power plants and, on the other hand, moderate tariffs for consumers (at the level of average power production, transportation, and distribution costs throughout the power system). In parallel with the central planning of the main indices and proportions of social and economic development of the country, this made it possible to effectively manage the widely expanded power economy and more thoroughly use the initiatives and capabilities of regional bodies and individual companies.

Box 4 Electric Power Industry in Planned and Market Economies

1. The regulated power monopolies (Model 1) and the regulated single-buyer market (Model 2) have a certain similarity with conditions of the planned economy. However, there are some distinctions in the mechanisms of EPS development planning and financing as well as price formation. The electricity tariffs for the indicated market models, in particular, will be individual for each regulated company, whereas in the planned economy they can be common for the whole country.
   Conditions for the competitive markets (Models 3 and 4) differ radically.
2. The state regulation of power companies is a rather sophisticated form of activity which has specific features in each country and should be constantly improved. In Russia, in the early 1990s, it was organized urgently from the outset. In China, transition from the state planning to the power industry regulation was performed gradually from 1997 to 2002.
3. In different countries, there are both private and state (corporate) generation companies. Despite the widely spread opinion that the activity of private companies is more efficient, the state companies have certain advantages:

   (a) The costs do not include "normal" profit intended for shareholders' dividends.
   (b) The profit gained is used in the State's interests.
   (c) Wages of managing staff, workers, and employees of the company are regulated.

4. In the centrally planned economy, the UPS of the USSR, being 100% state-owned, was the most reliable and cost-effective power interconnection. It enjoyed the effects of integrating interconnected power systems and regional power systems: decrease in generation capacities owing to diversity in peak loads reached 14 GW and the saving of emergency and maintenance reserves to 3–4%. The total effect in generation was 1.5–2.5 times higher than the costs for expansion of the UPS backbone network.

Chapter 3
Electric Power Industry in the Context of Microeconomics

This chapter analyzes the electric power industry in terms of the theory of microeconomics to show the distinctions of the electricity market from markets of other industries. It presents basic notions of microeconomics needed for such an analysis (Sect. 3.1) and a general idea about market types that are traditionally treated in microeconomics and their interpretation as applied to the electric power industry (Sect. 3.2).

3.1 Basic Notions of Microeconomics

Macroeconomics and microeconomics are usually distinguished in the theory of economics that studies primarily market economy [34, 40]. There is no clear border between them, and they are interrelated in many ways. However, there are certain differences between them.

Macroeconomics studies the economy of a country as a whole or its aggregated components (state and private sectors, industrial complex, sphere of services, etc.) and also international economic relations. It manipulates such economic indices as *total* production volume, *general* level of employment, *total* revenues and expenditures, and *general* price level.

The subject of *microeconomics* includes individual industries, firms (companies), and households. It studies production volumes and prices of a *specific* product, revenues, expenditures, and profit of a *separate* firm or company, and the markets of goods and services in *one* industry.

Figuratively speaking, macroeconomics studies the forest, and microeconomics the trees [34].

Market in the electric power industry is a subject of microeconomics, whose basic notions are discussed below. To begin with, we present some definitions borrowed from [34]:

- *Industry* is a group of *firms* (one or more) that produce identical or similar products.

L. S. Belyaev, *Electricity Market Reforms*,
DOI 10.1007/978-1-4419-5612-5_3, © Springer Science+Business Media, LLC 2011

- *Firm* is an organization that utilizes resources to produce goods or services for getting profit and owns one or several *plants*.
- *Plant is a real* enterprise (including earth and capital) that performs one or more functions in production, manufacture, and sale of goods and services.
- *Household* is an economic unit that consists of one or more individuals providing the economy with resources and using revenues received to purchase goods and services satisfying man's material needs.
- *Service* is an intangible process for which a customer, firm, or government is ready to pay.
- *Standardized product* is a product, all units of which are completely interchangeable and hence identical.
- *Market* is an institute or a mechanism that brings buyers (demand bearers) and sellers (suppliers) of a particular good or service together.

As to the market in the electric power industry, the *industry* is actually an electric power system (EPS) covering a certain area by its networks (the territorial limitedness of electricity market was considered in Sect. 2.3). In the National or Unified EPS of the country it will be one industry (and market). If the country has several independent EPSs or IPSs (interconnected EPSs), then we should consider correspondingly several "electric power industries" and their respective electricity markets.

The *firms* will be represented by individual companies that can be vertically integrated, power-generating, network or sales companies depending on the type (model) of electricity market organization. The vertically integrated companies and the majority of power-generating companies will own *several* (many) *plants—power plants*. This fact, as Chap. 5 shows, greatly influences price formation in electricity markets. The network companies (when the spheres of electricity transmission and distribution are separated) will perform, in fact, *service* functions rendering services on electricity transportation. The sales companies are resellers in the retail electricity markets.

In microeconomics the notion of *household* is rather broad. It includes not only the population furnishing one of the major production resources for firms—manpower (or simply "labor"). It also incorporates owners of two important resources—earth and capital. The "earth" in this case may be understood as minerals as well. When speaking about the electric power industry, the households are also electricity consumers. Besides, electricity is consumed by other industries, part of which, in turn, supplies energy companies with the required materials and equipment.

As was mentioned in Sect. 2.1, electricity is a *standardized* product, which strongly influences organization, operation, and expansion of the electricity market.

In any market, there are naturally *buyers* (or *consumers*) and *sellers* (or *producers*)[1] of corresponding goods and services. The buyers enter the market with their *demand* and the sellers with their *supply*.

---

[1] The terms "buyer" and "consumer" and also "seller" and "producer" will be applied as synonyms with the exception of specially mentioned cases.

Demand is characterized by a curve showing how much goods or services the buyer (or buyers) is ready to purchase at different prices at a certain period of time. The laid-down economic *law of demand* reads that all other things equal, the rise in the price of goods lowers their demand and, vice versa, the price reduction raises their demand. This fact determines a slope of demand curve in the coordinates "price–volume of demand" (as a rule, the price is plotted as an ordinate, and the volume of demand or supply as an abscissa). Two extreme cases are possible: (1) *perfectly elastic* demand, when in the indicated coordinates the demand curve is a *horizontal* line (at the fixed price) and (2) *perfectly inelastic* demand, when the demand curve is a *vertical* line (at the fixed volume of demand). The demand will be denoted by *D*, as is accepted in the literature on microeconomics.

Supply is described by the curve that shows the volume of goods or services to be offered by the seller (or sellers) at different prices during a certain period. The known corresponding *law of supply* reads that all other things equal, the rise of product price involves an increase in supply and vice versa. Therefore, the slope of the supply curve will be opposite to the slope of the demand curve (in the same coordinates). Similar to demand, there are two extreme cases—*perfectly elastic and perfectly inelastic* supply. In the first case the supply curve will be a horizontal line and in the second a vertical one. Supply will be denoted by *S*.

In the theory of microeconomics, the situations arising in the market are usually illustrated by *graphics*. In the free markets (without price regulation or manipulation) the equilibrium determining price and volume of sales (purchase) of goods is attained at the intersection of the curves of demand (*D*) and supply (*S*), as is shown in Fig. 3.1. The price p_e and the volume Q_e will be the equilibrium market price and the equilibrium volume of sales (purchase), respectively.

One of the key notions in microeconomics is *price*, i.e., the amount of money paid or received for a unit of good, service, or resource. The product price and its purchase volume specify the buyer's *expenditures*, and the price and volume of sales specify the seller's *revenues*. In some types of market, the prices can be regulated by the state, the regional, or the municipal bodies. The *regulated prices*, as was already noted, will be called *tariffs* in the book so as to underline their difference from market prices.

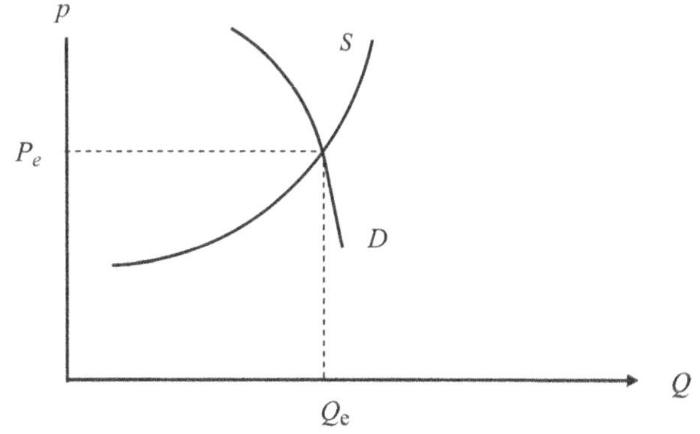

Fig. 3.1 Market equilibrium

3.1.1 Production Cost

One more important notion to deal with in greater detail is *cost*, i.e., the firm's expenditures for production of goods or services sold during a certain period of time. Production cost is a very versatile notion, and microeconomics addresses different types of cost.

It is common practice, first of all, to distinguish *explicit* and *implicit* costs. *Explicit* cost is direct *cash* expenditures of the firm for payment of labor force and transport services, purchase of raw material, fuel, etc. As a rule, this cost is taken into account in book-keeping. *Implicit* cost is unpaid expenditures of the firm's *own* resources. Two basic components of implicit cost are worth to be noted. They are *depreciation of buildings and equipment* that belong to the firm and *normal profit*. Depreciation does not apparently require special explanations, but the second component needs particular consideration.

Normal profit is believed to be an obligatory component of cost, since without it no firm can normally function and will shut down in the course of time. In [34], normal profit is interpreted as a *minimum payment for the entrepreneurial ability that encourages its use in the firm's business*. In principle, it may also be understood as a source of dividend payment to stockholders of the firm (equal to the minimum amount, below which the stockholders will start selling their shares). In the literature the specific value of normal profit is not usually indicated. Supposedly, it lies within 2–5% per year of the invested (or available) capital. Below we will assume that the normal profit is *always* included in the production cost.

The firm naturally can (and tends to) get profits in excess of the normal one. Such *extra profits* will be considered later.

In microeconomics the cost includes all payments and expenditures (both explicit and implicit) of the firm, including the normal profit. All of them are necessary to attract resources to a specific production sphere and to keep them in. In this book we will adhere to such a concept of cost.

Before we pass to a further discussion of cost, it is necessary to dwell on the very important notions of *short* and *long runs*. Production cost for these periods has a different sense. The market types also change.

In microeconomics the *short run* is an interval of time during which the *firm's production capacities remain invariable* (*fixed*). The volumes of product output and correspondingly the rate of capacity use can surely change. But the maximum possible product output is limited by these fixed production capacities. In the electric power industry, the term refers to fixed installed capacities of *operating* power plants that can work for different numbers of hours, i.e., generate different amounts of electricity. Naturally, there will exist a maximum possible volume of electricity production (e.g., for a year).

The *long run* is a period during which the *firm's production capacities change*. From here on, we will address the development (increase) of the firm's production capacities as a more general and interesting case. In so doing, the firm naturally bears a new type of expenditure—investments in the creation of new (or expansion of existing) capacities. In the long run, EPSs in the power industry will develop by construction of new power plants and transmission lines.

3.1.2 The Firm's Costs in the Short Run

The total costs for a certain period (e.g., for a year) consist of fixed and variable costs. The value of *fixed costs* does not depend on the production volumes. They include depreciation charges, lease payments, insurance premiums, interests on loans, wages of the permanent staff, and some other expenditures of the firm. The *variable costs*, on the contrary, vary with the change in the production volumes. They include expenditures on raw material, fuel, transport services, a certain part of labor, and similar variable resources.

For convenience, we will use the notation of different quantities that is accepted in microeconomics (several Latin letters). Then the costs for the short run can be written as below:

$$TC = FC + VC, \tag{3.1}$$

where the total cost (TC), fixed cost (FC), and variable cost (VC) are measured in rubles or dollars spent for the considered period (for example, rub./year or $/year).

Microeconomics most frequently deals with *average costs* (per unit of product). In this case, the full costs for the short run are divided in expression (3.1) by the volume of production Q at this period. The average costs will be expressed in the same units as the prices of production (rub./t, $/t, rub./kWh, etc.). However, the relations between such costs and production volumes prove to be complex enough.

Figure 3.2 borrowed from [34] presents a fundamental form of such relations for the average costs:

- The average fixed costs (AFC) that decrease with the increasing production volume (the firm's fixed costs are divided by the ever-increasing volume of production).

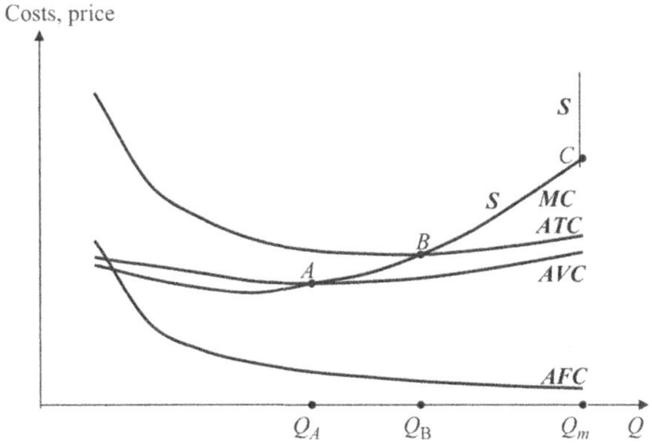

Fig. 3.2 The relationship between average fixed, variable, total costs and marginal costs

- The average variable costs (*AVC*) that reach their minimum (point *A*) at a certain production volume. The concave downward shape of the curve for *AVC* is caused by the *law of diminishing marginal returns*, which will be discussed separately.
- The average total costs (*ATC*) that are equal to the sum (vertically) of fixed and variable costs ($ATC= AFC+AVC$). The curve *ATC* has also a minimum, however, at somewhat larger production volume (point *B*).

The law of diminishing marginal returns reads that if the quantities of some factors are fixed, the marginal product of any variable factor (e.g., labor) will diminish with the increasing expenditure of this factor upon reaching some throughput [40]. Though the Law imparts a general character, it is illustrated and acts mainly in respect of *labor efficiency*, i.e., decrease in its productivity with intensification. Hence, the U-shaped form of the curves of average variable costs (*AVC*) is true in the case of considerable labor use by the firm to produce an output.

Meanwhile, because of high mechanization and automation of production processes—the EPS property that was indicated in Sect. 2.3—the labor inputs at power plants are weakly dependent or even completely independent of the amount of electricity produced. As will be shown in Sect. 5.3, the shape of curves for the average costs of power plants differs drastically from that in Fig. 3.2, making the electricity market very particular.

Note that all the curves in Fig. 3.2 are plotted for the *short run*, when the firm's production capacities are fixed. The maximum possible production volume of the firm is denoted by Q_m.

3.1.3 Marginal Costs and the Supply Curve of the Firm

Figure 3.2 presents a curve of *marginal costs* (*MC*) that characterize an *increase* in the costs for output of an *additional unit* of product. Whereas the above considered average costs were determined by division of the corresponding costs of the firm for a certain period by the total production volume for the same period, the marginal costs for each production volume *Q* are the differential:

$$MC = \frac{dTC}{dQ} = \frac{dVC}{dQ}, \tag{3.2}$$

where *TC* is the total cost and *VC* the variable cost for the production volume *Q*.

The value of fixed costs does not naturally influence marginal costs and solely the variable costs rise for the output of an additional unit of the product.

The curve of marginal costs (*MC*) cuts the curves of average total costs (*ATC*) and average variable costs (*AVC*) at the points of the minimum value of each (at points *B* and *A*, respectively). When *MC* are lower than *ATC*, the latter decrease, and when *MC* are higher than *ATC*, the latter increase. Likewise, when *MC* are lower than *AVC*, the latter decrease, and when they are higher, the latter, on the contrary, rise.

The marginal costs play a very important role when the firm participates in the market, where it sells its products (or services) at some price p. In the theory of microeconomics [34, 40, 41], it is shown that the firm will gain the maximum profit at the production volume that makes marginal costs equal to the price of products sold:

$$MC = p. \tag{3.3}$$

As Fig. 3.2 shows, after intersection with the curve AVC (point A), the marginal costs rise because of the law of diminishing marginal returns. Without going into detail, we will note the following features:

1. At the market price p that is *below* the value of marginal costs at point A, the firm cannot recover even its *variable* costs. If such a low price will be maintained for long, the firm should shut down.
2. At the prices between points A and B, the firm recovers *fixed* costs incompletely and cannot also function for long.
3. At the market price equal to the marginal costs at point B, it is profitable for the firm to produce the corresponding output Q_B. In this case it will fully recover the average total costs including the *normal* profit.
4. At the market prices *above* point B, the firm will start gaining an additional profit, the so-called *economic* profit. This profit per unit of product equals the difference between the market price and the average total costs ($p - ATC$), and its absolute value is equal to the product $(p - ATC) \times Q$, i.e., will depend on production volume.

In order to get the maximum economic profit, the firm has to set the production volume in accordance with equality (3.3). At the price p, it just recovers additional (marginal) costs for output of an additional unit of product. Hence, at the prices above point A, the firm should appear in the market with the offer of product output in accordance with the curve MC, which will be the firm's *supply curve.*

It should be noted that on the section A–B, the firm also has to follow equality (3.3). In this case *its losses will be minimized.*

Thus, the curve of marginal costs (MC) on the section A–B–C (up to the firm's maximum capacity Q_m) is simultaneously the curve of the firm's supply (S). Note that earlier in Fig. 3.1, the market equilibrium was shown for the whole industry and here the curve of supply of an individual firm is dealt with.

When the firm's maximum capacity Q_m (point C) is reached, its curve of supply (S) becomes a vertical range, i.e., *the firm's supply becomes perfectly inelastic*. At any market price above point C, the firm cannot output and supply more products than Q_m.

Let us consider now the *supply curve of the industry* that consists of several firms (for the whole market of this product). It is plotted by summing (horizontally) production volumes of individual firms that correspond to one and the same product price. It is natural, since at each value of the market price all the firms offer production volumes in accordance with their supply curves.

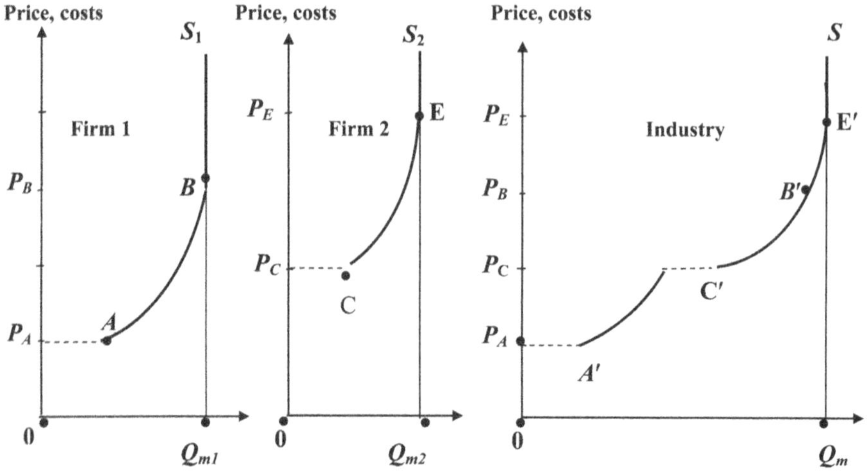

Fig. 3.3 Aggregate supply curve of the industry consisting of two firms

Figure 3.3, which is plotted on the basis of [40], presents an example of the industry consisting of two firms with the maximum capacities Q_{m1} and Q_{m2} (the supply curve of the industry with a greater number of firms will be plotted in a similar way). The points *A* and *C* are supposed to correspond to the minimum volumes of supply, with which the firms will enter the market. When the maximum production volumes are reached, the firms' supply curves (S_1 and S_2) become vertical ranges. The aggregate supply curve (*S*) of the industry is presented in the right part of the figure. In the range of prices P_A–P_C, only the first firm will participate in the market. Then, up to the price P_B, both firms enter the market. In this case, at the price P_C, the curve *S* will have a gap, after which it slopes down (the curves S_1 and S_2 are summed up). Above the price P_B, the first firm supplies the maximum possible volume Q_{m1} and the shape of the curve *S* is similar to the curve S_2 before the price P_E. At the price above P_E, the aggregate supply curve becomes vertical.

According to the aggregate supply curve *S*, the equilibrium market price and sales (purchase) volumes are formed (Fig. 3.1). The ways of constructing the demand curve *D* will not be considered. They will be specific for different types of products. We will suppose that they reflect the dependence of commodity or service purchase volumes on their price and are in one way or another expressed quantitatively.

One should bear in mind that the cost and supply curves of the firms and industry were considered above as applied to *the short run*, when production capacities of the firms are fixed. For the long run, when the firms (and the industry) develop, the sense of these indexes and the type of their dependences on production volumes will be essentially different.

The length of the short run, i.e., the period during which the production capacities of the firm do not change, can vary—a month, a quarter, a year, or even several years. However, it will be most natural to assume it to be equal to 1 year:

- The results of the economic activity of the firms, including power plants or power companies (costs, revenues, profit, etc.), are normally summed up for the whole year.
- The power industry is characterized by annual and intra-annual consumer load cyclicity. The variability of economic indices within a year will be appropriately reflected in the yearly indices.
- Under price regulation the electricity tariffs are usually revised once a year (taking into account inflation and other factors).

Further, for greater certainty and clarity, the *length of the short run will be assumed to be* 1 *year*. Reservations will be made when necessary. In this case, the production Q in Figs. 3.1, 3.2, and 3.3 will represent annual volumes, and costs or prices will be expressed in rubles or dollars per unit of product (for example \$/kWh).

The shape of cost curves that are shown in Fig. 3.2 is typical of the firms in the majority of sectors of the economy (which are considered in the theory of microeconomics) and supposes that:

- First, the average variable cost (*AVC*) curves and average total cost (*ATC*) curves have a *minimum* (their U-shaped form), which is due to the action of the *law of diminishing marginal returns*.
- Second, the supply curve of the firm (*S*) coincides with the curve of marginal costs (*MC*) at the section *above the minimum* of the *ATC* curve (point *B*), i.e., when marginal costs *rise above* the average total costs (*MC*>*ATC*). Hence, the firm enters the market with a supply curve that makes possible both—to *fully cover its total costs and to obtain its economic profit* (above normal profit).

In the power industry, most types of power plants have the short-run (yearly) cost curves essentially different from the same "typical" curves and their quantification is very difficult. Besides, the maximum electricity production (Q_m) at fixed installed capacities can depend on random natural factors (water inflow for hydropower plants, ambient air temperature for thermal power plants, etc.). This is considered in more detail in Chap. 5.

3.1.4 The Firm's Costs in the Long Run

As already mentioned, *long-run costs* are determined on the assumption that the production capacity of the firm is not fixed and can vary. Let us consider the case where the firm is developing, i.e., commodity production is increasing (though there can be an opposite situation).The long-run costs should include capital investments required to increase the productivity of the firm or company. Here, unlike the short-run costs, these costs are not divided into fixed and variable costs. Since capacity may vary, all factors or resources used by the firm can vary. Therefore, *the long-run costs*, in fact, have no *constant component and consist of variable costs only* (the total costs coincide with the variable costs).

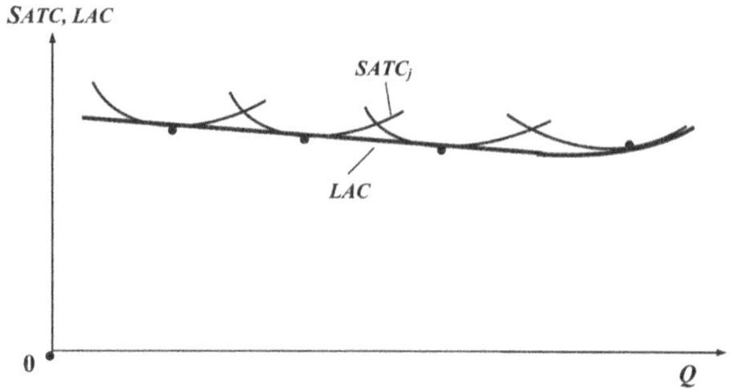

Fig. 3.4 The curve of long-run average costs

To illustrate the sense and method of constructing relationships between the long-run costs and production output, the theory of microeconomics [34, 40] assumes that the firm adopts an optimal policy of development. Several variants j of increasing the firm's productivity are considered and for each variant the short-run average total cost curve is constructed. In order to distinguish these two types of costs we will add the letter "S" to the short-run costs—$SATC_j$ and the letter "L" to the long-run costs—LAC. The curve of long-run average costs LAC will represent the lower envelope of these curves $SATC_j$ (Fig. 3.4).

Figure 3.4, constructed in accordance with [40], shows these curves. It is assumed that in the fixed component of $SATC_j$, for each jth variant of productivity Q_j the firm's capital investments in development to the level of this productivity are included. If we imagine that the total number of variants J is rather large, then the LAC curve, being an envelope, will turn into a smooth curve (and at some section even into a straight line). The downward slope of the LAC curve implies a decrease in costs as the firm's productivity increases, i.e., economies of scale. As was mentioned in Sect. 2.2, this effect is typical of EPSs. It has led to the formation of natural monopolies in the power industry. At the point where the LAC curve starts rising, economies of scale turn into diseconomies.

Along with the average costs there are long-run *marginal* costs (LMC), which are not shown in Fig. 3.4. At a descending section of LAC, the LMC curve will pass below LAC. At the point of the LAC curve minimum, they will intersect and at an ascending section of LAC the LMC curve will run above the LAC curve. The curve of long-run marginal costs is the *firm's supply curve* in the long run. Together with the curve of *long-run demand* of consumers (which is also not shown in the figure), it determines the point of market equilibrium in the long run (taking into account the firm's development). The demand of power consumers in the long run will be more elastic than the short-run demand, since there is a possibility to carry out energy saving measures, regulate load maximums, etc.

The market equilibrium in the long run based on the long-run demand and supply curves represents a "long-run" market in the power industry. This very market should be considered if the EPS expands, and the prices formed in this market should determine the electricity prices. These issues are considered in Chap. 6.

Box 5 Basic Notions of Microeconomics

1. The market in the power industry is a subject studied by *microeconomics* that deals with volumes of output and prices of specific products, revenues, costs and profit of individual firms, and markets of commodities and services within one industry. Buyers (consumers) participate in the market with the curves of *demand* (*D*), whereas sellers (producers) participate with the curves of *supply* (*S*).
2. It is very important to distinguish the *short run*, a period during which the firm's production capacities remain invariable (fixed) and the *long run*, a period during which the firm's capacities can change. For clarity, we will take 1 year as the length of the short-run period.
3. The "typical" firms considered in microeconomics are supposed to have a *U-shaped* (with minimum) form of relations between the *short-run* average costs and the product output. The supply curve of the firm (*S*) coincides with the curve of marginal costs (*MC*) after the minimum point of the average total cost (*ATC*) curve, i.e., when the inequality $MC > ATC$ is satisfied. Hence, the firm is supposed to enter the market with the supply that will fully cover its total costs and provide an economic profit. Meanwhile, as shown in Chap. 5, the cost curves of power plants have quite a different shape. This influences the supply curves, with which they may enter the market.
4. In the long run, the firm's costs include investments required to increase its production capacities. The long-run average costs will decrease at the economies of scale and, on the contrary, increase at the firm's diseconomies of scale.

3.2 Types of Markets for Commodities, Resources, and Services

In microeconomics [34, 40, 41] consideration is given to several types of markets:

1. Markets with *perfect* (*pure*) *competition* which will be called shortly *perfect* markets. Such markets are considered most effective and are taken as a reference (sample), though in reality they are quite rare (mostly in the agriculture). There are numerous conditions and requirements to be met in the market for the competition to be perfect: a great number of sellers and buyers, each being unable to affect the market price, their free access to the market and exit, etc.
2. *Absolute* (*pure*) *monopoly,* when there is *only one seller* in the market. This market is in absolute opposition to the previous one—an extreme case of an *absolutely uncompetitive* market. In particular, this monopoly can be observed in the power industry.

3. *Natural* (*regulated*) *monopoly*, which is effective if owing to economies of scale one firm in the industry can produce all the commodities at lower costs (and prices) than two or a larger number of firms. This situation, as was mentioned in Sect. 2.2, is characteristic of the power industry. In this case, the activity of firm and prices of products should be regulated by the state (regional, municipal) bodies for the firm not to abuse its monopoly position.
4. *Oligopoly*, when there are several sellers in the market and *entry of new sellers* into the market *is either complicated or impossible*. With "fair" competition, oligopoly can be very effective; however, there can be price manipulations—the use of market power by oligopolists, particularly under their collusion. The oligopoly situation is possible in the power industry if electricity prices are not regulated.
5. *Monopolistic competition* is typical of markets with partly interchangeable commodities (e.g., cars) which vary in quality and consumer properties. This kind of market is not characteristic of the power industry.
6. *Monopsony*, where there is only one buyer in the market. Here, unlike monopoly, it is the buyer, not the seller, who is in a privileged position (possesses market power). In microeconomics, this situation is considered mainly as applied to the manufacturer of some commodity, i.e., a firm which is the only buyer of a certain resource required for its production. Most often this resource appears to be labor. Meanwhile, in the power industry the single-buyer market model is possible (as was mentioned in Sect. 2.3). This model implies that the sellers (many of them) will be producers of a ready product—electricity. However, the firm (power company) that performs the function of the "single buyer" (it is also called a "Purchasing Agency") will be a monopoly reseller for the final electricity consumers. Here, like in the case of natural monopoly, the state regulation of electricity prices is required.

Oligopsony is a kind of monopsony with several buyers in the market. This kind of market is seldom considered in the theory of microeconomics. The possibility of organizing such a market (regulated) in the power industry should not be excluded. For example, the electricity market that has emerged in the past years in Brazil (and is forming in Chile) represents in general the single-buyer market. Actually, there are several buyers there—distribution-sales territorial companies. Therefore, this market can be referred to as oligopsony as well.

There are also other types of markets (e.g., price discrimination) that are of no interest in relation to the power industry.

All markets, except for the first one (with perfect competition), are *imperfectly competitive* or simply *imperfect*.

The models of the electricity market organization that were mentioned in Sect. 2.3 and considered in more detail in Chap. 4 correspond in one way or another to the foregoing types of markets. Model 1 represents a regulated natural monopoly, Model 2 represents a monopsony (also regulated), and Models 3 and 4 suggest, in fact, the presence of conditions for perfect competition. Besides, in the absence of electricity price regulation (Models 3 and 4) the oligopoly can be formed.

In the following sections two types of market will be considered in more detail: the first one to show the extent to which the electricity market does not meet the conditions and requirements for perfect competition, and the second to illustrate the behavior of monopolists and the formation of monopoly profit when the prices are not regulated. The oligopoly situation will also be characterized briefly.

3.2.1 Markets with Perfect Competition

Many conditions *for perfect competition to emerge* (*to be provided*) have been formulated. In [40], for example, the authors point out five such conditions:

1. *Many sellers and buyers participate in the market. The share of each of them is small with respect to the entire market, and therefore they cannot affect the price* (the price does not depend on supply or demand of individual market participants). In [23], this condition is interpreted as price-taking suppliers and buyers, i.e., those not trying to increase or decrease the price.
2. *Goods are homogeneous,* i.e., meeting the established standards. Therefore, the buyers do not care which seller to choose.
3. *Buyers are well informed about the sellers' price*—any seller increasing price loses its customers.
4. *Buyers and sellers act independently of each other.* They do not participate in price collusions. Each firm chooses the output volume that maximizes its profit on the assumption that it cannot affect the price. The buyers choose the volume of purchases acceptable for them at a given price.
5. *The firms can freely enter and exit the industry.* This condition is taken to guarantee that the firms existing in the industry cannot increase the price through agreement about output reduction, since any price increase will attract new firms into the industry which will raise the supply volume.

In addition to the enumerated conditions, some works name other conditions for perfect competition. For example, in [23] the author gives two more conditions:

6. *A good shape of the firm's short-run cost curve* ("well-behaved costs")—short-run marginal costs start rising while average costs stop shrinking after the firm reaches a certain (not very large) volume of production (i.e., the U-shaped forms of *AVC* and *ATC* curves are provided as is shown in Fig. 3.2).
7. In the long run the *characteristics of production costs* of a firm *should not create conditions for natural monopoly*. This implies that the curve of long-run average costs (*LAC*) does not have a descending form, as is shown in Fig. 3.4, but on the contrary, an ascending one. In other words, there should be diseconomies of scale in the industry.

In [51], the author points out one more condition for perfect competition that was determined by Nobel Laureates Gerard Debreu and Kenneth Arrow:

8. *Every market participant can buy an insurance against any possible risk.* This condition can be considered rather important.

Analysis of these eight conditions *as applied to the electricity market* shows that only condition 2 (a homogeneous or standardized product) is met in full measure. But it should be noted that some market organization models in the power industry foresee, along with the electricity market, the creation of markets for capacity, ancillary services, derivatives, etc., i.e., markets for several "products." This makes the market in the electric power industry imperfect and more complicated.

Conditions 5 and 7 are not met in the power industry at all. *Free entry* of producers into the industry *is physically impossible*, because this will call for construction of a power plant and its connection to an EPS (this property of the EPS was considered in Sect. 2.3). A physical barrier for new electricity producers makes the market imperfect in the *short run* when the installed capacities of power plants are fixed. Besides, in some models of the electricity market organization (Models 3 and 4) the *economic entry barrier* is created in the *long run* (this was mentioned in Sect. 2.3 and will be considered in detail in Chap. 6). The economies of scale that fostered the formation of natural monopolies in the power industry, as was discussed in Sect. 2.2, are *typical of the EPS.* In this connection it should be admitted that the *electricity market is imperfect.* This circumstance, by the way, is not disputed by anyone. However, the efforts to introduce competition are persistently made either in the hope to overcome this imperfection or for other reasons.

It is very difficult to meet condition 3 (buyers know well the prices of sellers) in the electric power industry (and not only in this industry). This condition is considered to be particularly important. It is spoken of in all the works devoted to perfect competition. In [23], the author speaks about *adequate* information available for all market participants, in [51] about *perfect* information, etc. In 2001, Joseph Stiglitz was awarded the Nobel prize in economics for his theory of "information asymmetry," i.e., the demonstration of the fact that information is not equally distributed among the market participants (this is spoken of in [51]). Hence, we can consider that this condition for perfect competition is not met everywhere. The difficulties in providing adequate information in the electricity markets are considered in [4, 23] and in other works.

It may seem that conditions 1 and 4 can be met, since in large EPSs (and markets on their territory) there are many power plants and the number of power consumers is even larger. However, as a rule, power plants belong to a relatively small number of generating companies. In the competitive electricity markets with unregulated prices (Models 3 and 4), these companies retain *market power* to one extent or another and can even form an oligopoly. The examples of market power and its analysis are presented in many publications [9, 23, 53].

The short-run cost curves of power plants (condition 6) are studied in Chap. 5. Anticipating things, we have to note that the minimum point of average costs practically for all types of power plants is reached at their maximum annual output, i.e., the shape of these curves is not "good" (U-shaped).

Finally, to meet the last condition (insurance against any risk), it is necessary to create a special system of insurance.

Thus, we can state that the *electricity market is not a market with perfect competition*, and organization of free competition (electricity price deregulation) can lead to undesirable consequences.

If competition in the market is *imperfect*, then without state regulation this market will be a kind of imperfect market: a monopoly or an oligopoly with dominating (market power) *sellers* and a monopsony or an oligopsony with dominating *buyers*. In the electric power industry that has the features of a *natural monopoly*, market power obviously belongs to sellers (producers). Without regulation (under "spontaneous" market) this will lead, as the experience in the early twentieth century shows, to formation of a monopoly. However, if the generation sphere is forcedly split into several independent power-generating companies, then without regulation it will be an oligopoly. These two market types will be considered in the following sections.

3.2.2 Monopoly Market

We will consider this market as applied to the power industry for the short run.

Figure 3.5 shows the main economic characteristics that affect the formation of electricity prices in the entire power system for a *monopoly company*, when characteristics of the power system and company coincide. The figure is plotted with the use of [40] in the coordinates of electricity prices P ($/kWh) and annual electricity output Q (kWh). It shows:

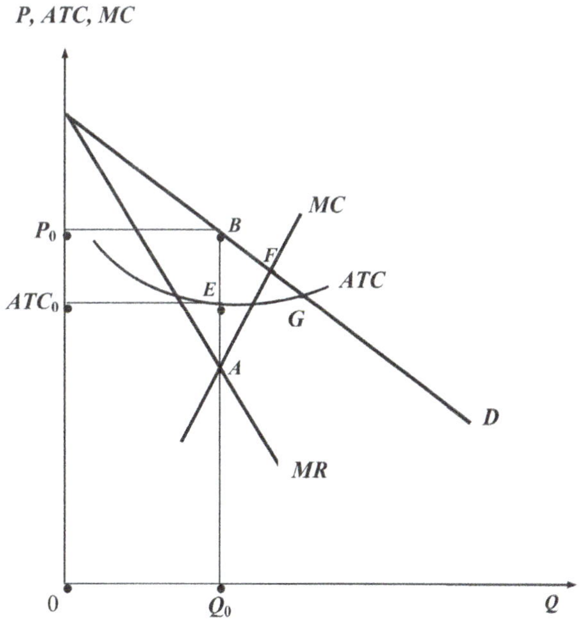

Fig. 3.5 Price formation in the wholesale electricity market under monopoly

- The dependence of short-run average total costs (*ATC*).
- The dependence of short-run marginal costs (*MC*).
- *A curve of demand* for electricity *D* (is shown as a line); it expresses annual volume of electricity that can be bought by consumers at a certain price *P*.
- The curve of the company's *marginal revenue* (*MR*) which characterizes the company's revenue increment at the annual output increase by 1 kWh; at the increase in annual output *Q* this specific marginal revenue decreases since the price at which consumers buy electricity drops; the *MR* curve depends on the curve of demand *D* and its slope is steeper than that of *D* (for more details see [34, 40, 41]).

The *MR* curve is very important for the economic activity of a monopoly company. It is as important as the *MC* curve is. The former expresses marginal (incremental) revenue, the latter marginal costs (per 1 kWh).When *MR* exceeds *MC*, it is profitable for the company to increase electricity production. Whereas if $MR < MC$, it is profitable for the company, on the contrary, to decrease production. Equality of these values

$$MR = MC \tag{3.4}$$

determines the company's optimal volume of production Q_0, at which it will get the *maximum profit*. In Fig. 3.5, it is reached at point *A* where curves *MR* and *MC* intersect.

If the monopoly company is *not regulated* and can arbitrarily vary its production volume, it will establish it in the amount of Q_0. This production volume will correspond to point *B* on the demand curve *D*, and the electricity price will rise to P_0. Meanwhile, average costs of the company at the output Q_0 will be at the level of ATC_0 (point *E* is somewhat higher than the point of the *ATC* curve minimum). Hence, the company will get *monopoly profit*[2] equal to the area of the rectangle EBP_0ATC_0. One can prove that if electricity production decreases or increases as compared to Q_0, this profit will be lower (see also [34, 40, 41]).

The conditions of market equilibrium (an abstract case when the company does not use its monopoly power) in Fig. 3.5 correspond to point *F* at which the curves *D* and *MC* intersect. Here, according to the theory of market microeconomics, the economic equilibrium is attained and the most comprehensive use of all available resources and factors (those of both producers and consumers) is provided. We can see that at point *F* electricity production (and consumption) will be higher and electricity prices will be lower than at point *B*. At the same time, at point *F*, the monopoly company will still receive extra profit (above the normal one), since its average total costs are still lower than the marginal ones (the *ATC* curve goes below the *MC* curve).

For *regulated natural monopolies*, tariffs are established at the level of average total costs. In Fig. 3.5 this corresponds to point *G* where *ATC* intersects the line *D*.

[2] In the general case, monopoly profit appears when for some reasons the prices in the market rise above the marginal costs of the firm [40]. This situation will be illustrated and explained further.

Product output here additionally increases and price drops as compared to point *F*. In this case the monopoly company gains only normal profit included in the production costs. This regulation is usually considered as a transfer of all advantages obtained from the natural monopoly to consumers. Indeed, under regulation, consumers get the largest amount of electricity and at the lowest prices.

Comparison of the three considered cases (points *B*, *F*, and *G* in Fig. 3.5) shows that:

- Termination of price regulation and creation of conditions for perfect competition in the wholesale market (transition from point *G* to *F*) will increase electricity prices from the level of average total costs to the level of marginal ones, which will result in decrease in electricity consumption.
- If the conditions for perfect competition are not provided, the producer can decrease electricity production for the price to exceed the marginal costs and the monopoly profit to be obtained (transition from point *F* to *B*).

3.2.3 Oligopoly

Organization of the wholesale electricity market with unregulated prices (Model 3) creates potential possibilities for the formation of an oligopoly, i.e., a situation when the number of sellers (power generation companies (PGCs)) in the market is not large. In the majority of the countries, the unbundling of the generation sphere of the monopoly vertically integrated company into several independent and unregulated PGCs represents a ready model of oligopoly. Normally, they seek to split generation into as many PGCs as possible, but in the electric power industry, this, on the one hand, leads to a decrease in economies of scale and, on the other hand, is not so important, since all generation companies (disregarding their number) have common (coinciding) interests.

The status, interests, and possibilities of generating companies that become independent differ considerably from the regulated monopoly vertically integrated companies:

- They are not responsible for reliable power supply to consumers.
- The maximum profit becomes their main interest.
- Risk associated with construction of new power plants is borne entirely by them (it is not transferred to consumers).
- Private (nongovernmental) companies can invest their available free funds (capital) into any sectors of the economy (not only into the power industry) if it brings larger profits.
- Despite competition with one another, they have common interests regarding consumers—to sell electricity at a higher price (to raise prices).
- They are not interested in new power producers (NPPs), since their appearance will increase supply and decrease prices.

Oligopolies are known to be dangerous in the case of collusion among oligopolists—they can increase the prices of their products. Most of the countries prosecute such collusions.

Meanwhile, in the electric power industry with physical and price barriers for the entry of NPPs into the market, there are possibilities to raise prices by *creating capacity and electricity shortage*, for example, by manipulation in the sphere of maintenance of units at power plants or electric tie lines, which was commonplace during the California crisis in 2000–2001.

However, the most radical method of creating shortage is to stop construction of new power plants, which will inevitably lead to power shortage and price rise as the electricity consumption increases. Here, the chances also are that oligopolists (existing PGCs) will construct power plants, but in an inadequate (insufficient) rate to maintain power shortage and high prices, yet not so high as to make the construction profitable for new producers (i.e., maintain price barriers to the entry of NPPs into the market). This possibility will be explained in more detail in Chap. 6.

Thus, with organization of the competitive wholesale market the *regulated* natural monopoly can turn into an *unregulated* oligopoly. The latter can also be considered "natural" due to the coincident interests of operating power companies and creation of oligopoly without their direct or apparent collusion. This possibility, in particular, is shown in [5, 19].

It should be noted that organization of the single-buyer market (Model 2) can prevent the creation of an oligopoly of electricity producers. This market will make it possible to implement the competition effect among producers and thus will deprive them of any market power. Such a "power," in this case, will belong to the purchaser (a single one). Producers will have to compete to sell him their electricity.

Monopsony in the electric power industry, as was already mentioned, should be regulated by the State (as well as monopoly). Under regulation, the company—a monopsonist—being in a privileged position with regard to producers, will in fact represent the interests of consumers, i.e., this model of electricity market organization is profitable for consumers.

Box 6 Possible Types of Markets in the Electric Power Industry

1. Among the markets studied in microeconomics, those of interest for the power industry are: a market with perfect competition, a natural regulated monopoly, an oligopoly, and a monopsony.
2. The electricity market does not satisfy virtually all (except for one or two) conditions of the *perfect* competition:

   (a) It is *physically impossible* for new producers to freely enter it and in some cases an *economic barrier* may exist.
   (b) The electricity buyers (consumers) lack *adequate* (full, timely) *information* about prices in the market, etc.

Organization of the competitive wholesale electricity market with free prices without conditions for perfect competition should be considered as theoretically groundless and open to many hazards.

3. The monopoly wholesale electricity market is characterized by three states, at which prices sequentially increase and consumer demand decreases:

   (a) Price regulation with its setting at the level of average total costs (*ATC*) of the monopoly company; the latter receives in this case only normal profit.
   (b) Creation of conditions for perfect competition (an abstract case), when the price is fixed at the level of marginal costs (*MC*) and the company receives an economic profit.
   (c) Complete deregulation, at which the company lowers production to the level of the equality for its marginal revenue and marginal costs ($MR = MC$) and gets a maximum monopoly profit.

4. The *oligopoly* with obvious and coincident interests of electricity producers (even without their collusion) can be formed in the competitive wholesale market. Therefore, transition from the regulated natural monopoly to the competitive market can prove to be a transition to the *unregulated* "natural" oligopoly.
5. The regulated *monopsony* (with respect to producers) can be a good type of market in the power industry. It implements the benefit of competition in the generation sphere, eliminates the use of market power by producers (in particular, oligopoly formation), brings electricity consumers into a privileged position, etc.

Chapter 4
Models of Electricity Market Organization

This Chapter deals with specific features of different electricity market models (Sect. 4.1) and makes their qualitative comparison in terms of several criteria, including analysis of the competition effects and their possible implementation with electricity price regulation (Sect. 4.2). The last Sect. 4.3 concisely enumerates the pitfalls of the competitive market.

4.1 Basic Models of Electricity Market Organization

In the market economy countries, the electricity market is inevitably organized in some way or another. It cannot be spontaneous because of peculiar features inherent in electric power systems (EPSs) and electricity significance. There are economic, natural, and social peculiarities of specific countries, which also add to a great variety of electricity markets. At the same time it is customary to distinguish four main models of such a market [32, 38] that have been mentioned earlier: a regulated natural monopoly, single buyer, competition in the wholesale market, and competition in the wholesale and retail markets. The markets of specific countries are modifications of one of these four markets.

4.1.1 Model 1—Regulated Natural Monopoly

Specialized electricity transportation and the economies of scale characteristic of EPSs make it possible to consider formation of the monopoly in electric power industry as most natural and probable. Spontaneous (uncontrolled or unorganized) electricity markets gradually became monopoly ones in the early twentieth century. After comprehension of this process and recognition of the monopoly electricity market advantages, the natural monopolies (vertically integrated companies) became institutionally legalized under the state regulation. Absence of regulation in the monopoly market would result in situations with limited production and price rise discussed in Sect. 3.2.

L. S. Belyaev, *Electricity Market Reforms*,
DOI 10.1007/978-1-4419-5612-5_4,

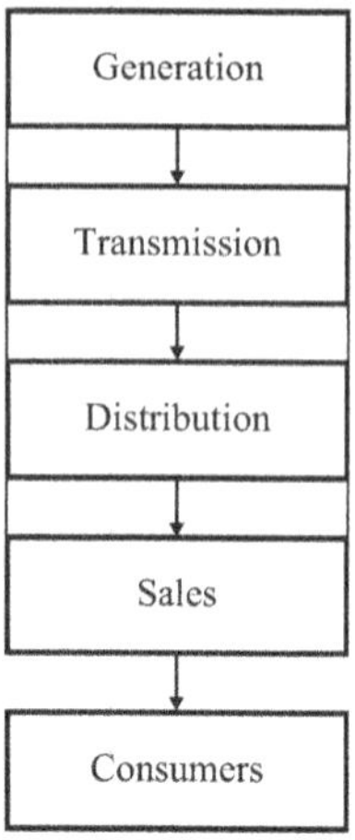

Fig. 4.1 Model 1—regulated monopoly

Vertically integrated monopoly companies cover all the spheres of electricity generation, transmission, distribution, and sale on their service territory (Fig. 4.1). They are charged with reliable (uninterrupted) electricity supply to consumers on their territory. *Independent power producers* (IPPs) and individual sales companies may exist or appear very often in parallel with monopoly companies. They sell or purchase electricity from the monopoly company by special agreements (also under the state control). Connection of IPPs to EPS networks owned by the regulated monopoly based on mutually beneficial terms is surely economically sound and reasonable. Starting in the 1970s it was legislatively settled almost in all countries with such a market model.

The consumers' prices (tariffs) for electricity produced by natural monopolies are fixed by the regulatory bodies at the level of average costs of the company (including normal profit). For the electric power industry with diverse types of power plants, this fact is of particular significance. Formation of tariffs based on the average costs means that the higher costs of less effective power plants (such as coal-fired CPPs) are compensated by lower costs of more effective power plants (such as HPPs). As a result a relatively low tariff for consumers is ensured (in the free competitive market the equilibrium prices will be set at the level of costs of the least effective power plants).

The monopoly companies are planning development of EPSs (both power plants and electric networks) on their service territory in terms of reliability and economic efficiency of power supply to consumers, environmental requirements, etc.

The plans on construction of new power plants and transmission lines (TLs) are coordinated with a regulatory body and then the expenditures on construction (capital investments) are included in the form of an investment component in electricity tariffs as the necessary expenses of the company. Thus, consumers guarantee payment of the costs for EPS development, even if the plans were not quite optimal or the company and the regulatory body were mistaken, for example, in the forecasts of electricity demand. Such a situation is commonly interpreted as the "shift of *risks* (impacts of uncertainty in future conditions or erroneous decisions) to consumers."

If new power plants and TLs are constructed on credits, then owing to their guaranteed repayment (absence of the risk) the monopoly company can obtain them at the low loan interests and the long term of repayment. Naturally, credit repayment with interests increases capital expenditures attributed to tariffs, but to a relatively small extent.

Thus, the key advantages of Model 1 are believed to be the following:

- The most complete capabilities for realization of the economies of scale (effects of creating and interconnecting EPSs that were discussed in Sect. 2.2).
- Low electricity tariffs for final consumers.
- Absence of EPS expansion problems.

The main disadvantage of Model 1 is thought to be the *cost-based principle of tariff formation*, i.e., inclusion of all expenditures of the company in tariffs that are recognized by the regulatory body as necessary and sound. In this case the company has no special incentives to enhance production efficiency and reduce capital expenditures for EPS expansion. As regards the latter, the private companies, on the contrary, have an incentive to overstate them for fixed capital. It was a reason of "overinvestment," i.e., accelerated commissioning of generation capacities in many countries and hence, creation of surplus capacity reserves (up to 30–40%), which is considered to be an additional disadvantage of regulated monopolies.

Regulatory bodies play a very important role in maintenance of effective operation and expansion of EPSs that belong to the monopoly companies. They should, therefore, be furnished with skilled specialists and supplied with proper guide documents, determining their rights, liabilities, and rules (procedures) of regulation. At the same time they should be protected from corruption, since they are subjected to constant pressure from the monopoly companies.

Along with the companies the regulatory bodies bear responsibility for reliable electricity supply to consumers. They are inclined, therefore, to have rather surplus generation capacities than their shortage. Such an inclination was another reason of the mentioned "overinvestment."

The indicated disadvantages and the increase in electricity tariffs in some countries gave rise to criticism of Model 1 in the late twentieth century and brought about proposals on restructuring electric power industry with implementation of competition in the areas of electricity generation and sale. The competition was expected to replace the state regulation, enhance production efficiency, and lead to electricity price reduction.

Meanwhile, there exists an alternative direction for enhancing power industry efficiency in Model 1—*improvement of the state regulation*. In particular, in [23] the author noted that the regulation in the United States has typically been passable enough and could be much better if the efforts spent deregulating were spent on improving regulation. One of the ways to improve regulation, namely to fix tariffs for a long term (several years), is considered in Sect. 4.2.

Model 1 is applied in the electric power industry of France, Japan, majority of states in the USA and provinces of Canada, and many developing countries.

4.1.2 Model 2—Single Buyer

This market model (Fig. 4.2) differs from the previous one by division of the generation sphere into several economically independent power generating companies (PGCs) that compete with one another for electricity supply to the single Purchasing Agency. Appearance of new power producers (NPPs) is also possible. The rest of spheres remain integrated within one company that is the monopolist with respect to consumers as before. This company (Purchasing Agency) should naturally continue to be regulated by the State.

In the microeconomics notions this model is an extremely sophisticated *imperfect* market. It combines elements of *the oligopoly* that can be formed by producers, *the monopsony* of the Purchasing Agency with respect to producers and its mentioned *monopoly* for consumers. It is rather difficult to imagine the work of such a market without the state regulation. The oligopolists (PGCs and NPPs) would decrease production volume to increase electricity prices. As the monopsonist the Purchasing Agency, on the one hand, would reduce the volume of purchases from producers to cut prices of electricity bought and, on the other hand, as the monopolist it would decrease the volume of sales to consumers to raise prices of electricity sold. Such manipulations would finally lead to increase in prices and losses for consumers.

The state regulation radically changes the situation. Regulation excludes the use of market power by the oligopolists, monopsonist, or monopolist. As it is shown below, Model 2 with regulated electricity prices is in some ways the best for consumers.

The Purchasing Agency remains responsible for uninterrupted supply of consumers. Therefore, it (like the regulated monopoly) must plan and develop EPSs on its territory in advance in order to avoid electricity shortage.

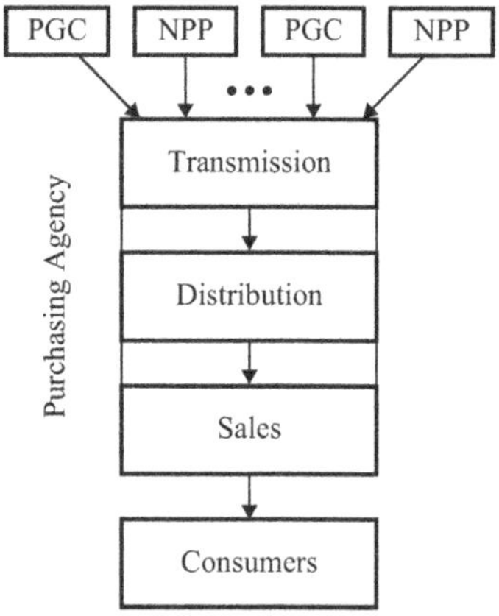

Fig. 4.2 Model 2—Single Buyer

The Purchasing Agency buys electricity from PGCs and NPPs on the basis of *long-term contracts* at the defined prices, date, and terms of delivery. The contracts with the operating producers are concluded for a period of 1–5 years. The prices are fixed for each operating producer individually at a level close to its production costs, including fixed and variable costs of power plants and also normal profit. The contracts with new producers are made for the period of 10–15 years that is sufficient for the payback of investments in a new power plant. In this case the electricity prices are set at a higher level than for the operating power plants of the same type.

The terms of supplies and particularly the prices settled in contracts of the Purchasing Agency with producers must be controlled by the regulatory bodies to avoid abuses to the detriment of consumers. Actually the prices (tariffs) will be regulated as for the regulated monopoly.

In addition, the long-term nature of contracts and their conclusion by producers with the same Purchasing Agency create a body of favorable opportunities. Firstly, the generation capacity excess makes the competing producers offer the lowest possible prices, thus the competition effect is realized. At the same time, if the contract is made for several years, the producers will have an incentive and time to lower production costs and gain a higher profit. Secondly, it is possible to prevent shortage of capacity and electricity. The Purchasing Agency is planning EPS expansion in advance and hence will forecast electricity consumption, draw up prospective balances of power, etc. It can conclude long-term contracts with PGCs or NPPs for additional electricity supplies (from new power plants) should the need arise. The term of these contracts should exceed the time required for pay back of capital investments in new power plants at the supplied electricity prices that are defined in the contract. Thus, the investor will have a guarantee of investment payback, which makes it possible to provide for a low interest rate in the contracts. The long-term contracts decrease risks and enhance financial stability of generating companies. Peculiarities of their conclusion have been analyzed in [32].

For the whole Purchasing Agency, the prices of different producers will be averaged in much the same way as it takes place in the regulated monopoly. Higher prices in the contracts on electricity supply from new power plants will also be averaged and hence final consumers will have a low level of tariffs.

The model single buyer realizes the competition effect among electricity producers. The generation costs are known to prevail (50–60%) in the full costs in power industry. Therefore, application of this market model will enable realization of the main part of the potential effect of competition, on the one hand, and gradually lead to decrease in production costs and tariffs for final consumers in comparison with the regulated monopoly subject to proper regulation, on the other hand. This is its principal advantage over the latter.

In addition, in this market model electricity consumers are in a "privileged" position with respect to producers. The Purchasing Agency (regulated) and the consumers dominate in the wholesale market. Electricity producers are deprived of any market power.

The shortcomings of Model 2 are usually tied with the necessary state regulation (as in Model 1) that encounters the mentioned difficulties. However, regulation of the wholesale prices has positive aspects for electricity consumers (that were already considered and will be considered later), on the one hand, and happens to be inevitable, e.g., at power shortage, on the other hand. Therefore, the regulation methods, rules, and procedures undoubtedly require continuous improvement.

The model Single Buyer has been realized now in the power industry of China, Republic of Korea, several countries of Latin America, and also in some other countries. In Russia this model was applied to organize the Federal wholesale market of electricity and capacity (FOREM) in the 1990s. Specific forms of drawing producers into competition differ depending on specific features of the country. In particular, they are substantially different in China, Republic of Korea, and Brazil (see Chap. 7). The common features (attributes) of the model Single Buyer seem to be organization of competition only among electricity producers, electricity price regulation, averaging of tariffs of different producers, and use of long-term contracts.

4.1.3 Model 3—Competition in the Wholesale Market

Model 3 (Fig. 4.3) considerably differs from the previous market model. Several distribution sales companies (DSCs) appear instead of one Purchasing Agency. These companies, as a rule, do not own generation capacities and monopolistically deliver electricity to consumers on their service territory. They bear responsibility for reliable power supply and remain regulated by the regional or municipal bodies (energy commissions), in particular concerning tariffs for electricity supplied to consumers. They possess low-voltage distribution networks that should be developed by them when needed.

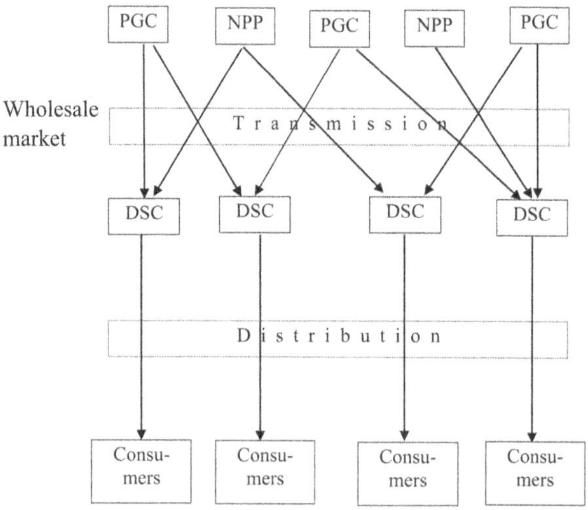

Fig. 4.3 Model 3—competition in the wholesale market

Following are new organizations (not shown in Fig. 4.3) that have emerged in the wholesale market:

- The transport network company (TNC) that owns high-voltage networks and delivers electricity from producers to DSCs. It is responsible for the free (but paid) access to its network for any electricity producers and buyers in the wholesale market.
- The independent System Operator (SO) that provides for dispatching control of electricity production and transportation.
- The independent Trading System Administrator (TSA) that arranges trade in electricity.

Trade in the wholesale electricity market can be organized at the free prices either through the *spot* market or by the bilateral *long-term contracts* (for 1–3 years) between producers and DSCs. The trade in the spot market is carried on in time close to real (with hourly or half-hourly intervals) by bids of producers and purchasers, with indication of the volumes and prices of sales or purchases for each hour of a day. The bids are submitted 1 day in advance and such a market is called the "day-ahead market." The prices of the spot market are set as equilibrium ones providing the balance between supply and demand within the corresponding hour of a day.

Note that the spot market is not really a *short-term* market in the sense it is interpreted in the theory of microeconomics (see Sect. 3.1) since the producers enter the market only with *variable* (moreover, hourly) costs. Only the market of long-term contracts, i.e., "forwards" will be a real short-term market. Its prices include the producers' *total* costs including the fixed ones. Meanwhile, the prices in the bilateral contracts are, as a rule, confidential, i.e., are known only to the companies making them. In this case, on the one hand, one of the key conditions of perfect competition, namely good information awareness about the sellers' prices, is violated. On the other hand, confidentiality of prices will lead to the fact that the market will not give in general any signals on the volumes of sales and purchases and the market expansion (or shrinkage). Hence, in the power industry it proves to be impossible to organize a "normal" market similar to the markets in other branches. This aspect is addressed in greater detail in Chap. 5.

In addition to the indicated markets in some countries with the spot markets, there are markets of ancillary services (to ensure reserves, frequency, voltage regulation, etc.), capacity markets, and markets of "derivatives" (futures, options). This fact considerably complicates electricity trade, on the one hand, and creates opportunities for diverse manipulations by producers to gain excess profits, on the other hand.

In Model 3, the sphere of electricity transmission is considered to be monopolistic. TNC, therefore, should be regulated by the state body concerning tariffs for using networks, their adequate development, and "normal" profitability support. TNC is responsible for the development of high-voltage networks, in particular provision of electricity trade. Investments in construction of new transmission lines that are coordinated with the regulatory body are naturally included in tariffs for access to the networks.

Models 1 and 2 also have bodies performing functions of dispatching control; however, these are included in the integrated companies. The SO activity in this model (as in Model 4) is greatly complicated by the necessary consideration of contractual electricity deliveries. This is particularly typical of the economically optimal loading of power plants in the process of change in daily load of consumers and of regulation in the overloaded network branches that are jointly used by different producers and purchasers.

Formally, the advantage of Model 3 is assumed to be the appearance of the wholesale market [32] with competition of both electricity producers and purchasers (DSCs), with the latter having an opportunity to choose a supplier. However, should we consider the situation carefully, it will be revealed that the competition among purchasers:

- Does not enhance production efficiency and does not reduce costs (as distinct from the competition among electricity producers).
- Deprives consumers of the privileged position, in which they are in Model 2.
- Sharply improves the position of producers that become deregulated and gain a market power over consumers (in practice it means price rise in the wholesale market to the equilibrium ones corresponding to costs of the most expensive power plants).

In parallel with structural changes, transition to Model 3 leads to radical changes in conditions, incentives, and mechanisms of financing *construction of new power plants.*

First, now there is no body (company or state commission) responsible for the development of generation capacities and deficit prevention in the wholesale electricity market. As it was noted, TNC bears responsibility only for the development of high-voltage networks. Although DSCs are formally responsible for electricity supply to consumers on their territory, in fact, they will not be able to provide it, if there is power shortage in the wholesale market. In case of the shortage, some DSCs will not be able to buy electricity even at very high prices.

Second, the risks of power plant construction are not shifted to consumers any longer (as it was typical of Models 1 and 2), but should be accepted by producers (or investors). This will increase the capital value, i.e., an interest rate, at which the investor will decide to construct a power plant. Besides, the investments in a new power plant should be recovered now by selling the electricity produced only at this power plant (in Models 1 and 2 these investments are recouped by all the EPS output), which creates a price (economic, financial) barrier for the new power producers to enter the wholesale market and can lead to formation of the oligopoly of existing producers, power shortage, and price rise (see Chap. 6).

Introduction of free competition in the wholesale market, *when there are no conditions for perfect competition in electric power industry,* is a general disadvantage of Model 3 (as was shown in Sect.3.2). Therefore, it could be expected (and practical experience confirms it) that introduction of the competitive wholesale market would lead to negative consequences. The revealed great number of specific shortcomings and negative properties of such a market will be indicated in the next section and analyzed in the later chapters of the book.

4.1.4 Model 4—Competition in the Wholesale and Retail Markets

Model 4 offers the possibility of competition in retail markets in addition to Model 3 (Fig. 4.4). Electricity consumers can be supplied now with electricity from different distribution sales companies (DSCs) or sales companies (SCs). The latter are a new participant of the market that appears in Model 4. They are only electricity resellers and have no own distribution networks. The remaining DSCs must provide a free access to their networks (for a certain charge) for any SCs and also producers (PGCs and NPPs) for electricity sale to consumers connected to their networks. Since electricity distribution remains a monopoly sphere, DSCs continue to be regulated by the state (regional, municipal) bodies as regards fixation of the payment for distribution network access and inclusion of expenditures for network development in it.

The most complete realization of Model 4 leads to creation of individual network (monopoly) companies responsible for maintenance and development of distribution networks and obliged to provide an access for any consumers, sales companies, and electricity producers to them. Consumers in this case buy electricity only from SCs or directly in the wholesale market.

Model 4 includes direct electricity supply from producers to consumers without DSC or SC. It means either direct appearance of consumers in the wholesale market or some combination (integration) of the wholesale and retail markets. In the latter case the procedures of accounting, mutual settlements, etc., are naturally complicated.

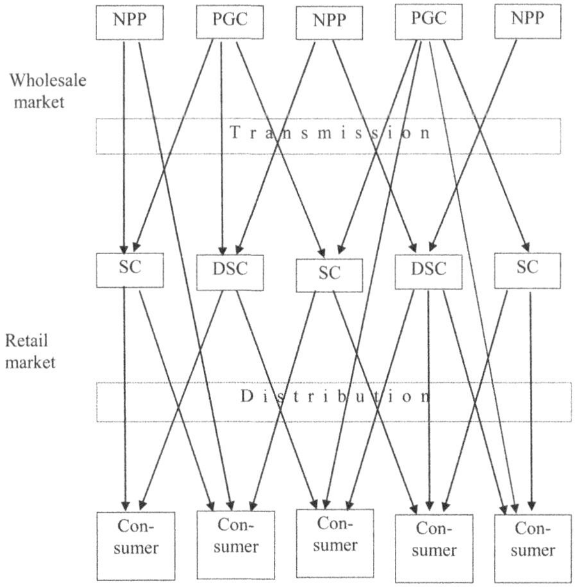

Fig. 4.4 Model 4—competition in the retail market

In the wholesale electricity market in Model 4, the organizations considered in Model 3, i.e., transport network company, independent System Operator, Trading System Administrator, etc., remain. Sharp increase in the number of buyers also complicates activities of these organizations.

Model 4 allows the consumers to choose suppliers, presenting thus some advantages in comparison with Model 3:

- The monopoly DSCs had no special incentives to buy the cheapest electricity, since in any case its cost was passed on to consumers; now SCs while competing with one another to attract consumers will make efforts to do this (in particular, in the bilateral contracts).
- The consumers' response to prices of the wholesale market increases (meaning the change in electricity demand or saving). In Model 3 it was partially damped by DSCs.
- DSCs could act in a nonoptimal way concerning logistics and other expenditures, since they were also shifted to consumers.

At the same time the problem of metering and accounting of electricity flows becomes very pressing in Model 4. Lots of consumers (all, in the limit) participate here in retail trade. In 1998 the consumers in Great Britain, e.g., numbered 22 million [32]. Such systems of accounting (and subsequent payments) can be created only by economically developed countries (in Chile and Brazil, e.g., the retail markets were not organized). These systems of accounting involve heavy expenditures that should be considered jointly with administrative charges of numerous sales companies while assessing efficiency of creating competitive retail electricity markets. The more so as the costs in the sale sphere are relatively low and account for about 5% of the overall costs in EPS. Their decrease due to competition can result in a minor effect that will hardly exceed expenditures on organization and operation of retail markets. The most probable situation is when consumers gain the possibility for choosing a supplier, but the electricity prices with any of them will be higher than they were earlier in the case of the monopoly supplier.

The problems concerning construction of new power plants that arose in Model 3 remain and are even aggravated. As before, there is no body responsible for the development of generation capacities and prevention of shortage in the wholesale electricity market. Besides, both the retail and wholesale markets are not perfect and this fact increases possible adverse consequences of electricity price deregulation.

Box 7 Principal Models of Market Organization in the Electric Power Industry

Model 1—regulated natural monopoly (a vertically integrated company). This model makes it possible to realize the economies of scale inherent in EPSs most completely, impose tariffs for consumers at the level of average costs of the monopoly company, ensure required EPS development. As a rule, there are also independent power producers and sales companies, whose connection to electric networks of the monopoly company with conclusion of appropriate contracts is advisable.

The main disadvantage of this (and the following) model is the necessary state regulation associated with certain problems.

Model 2—single buyer (Purchasing Agency), when only producers compete with one another in the wholesale market, and electricity is bought by the long-term contracts. The prices of electricity that is bought from producers and sold to consumers are state-regulated. The model implements the effect of competition in generation, ensures optimal development of generation capacities and low prices for electricity consumers that are in a "privileged" position.

Deregulation of prices and further unbundling and splitting of the spheres of electricity distribution and sale in the following models are a *pivotal reform* that creates the producers' market power with the consequences involved. Transition from Model 2 to Model 3 does not lead, in particular, to the cost decrease in the EPS generation sphere. The alternative way to enhance power industry efficiency is *to improve state regulation.*

Model 3—competition in the wholesale market with several electricity producers and purchasers; the latter exclusively distribute (resale) electricity to consumers on their service territory. The wholesale prices are not regulated and established (when there is no deficit) at the level of costs of the least effective power plants participating in the market. Because of general imperfection of the electricity market this model causes a set of problems that are considered further in the book.

Model 4—competition in the wholesale and retail markets, when both electricity producers and consumers freely compete with one another. Organization of the retail markets that cannot also be considered perfect requires complex systems to be created for accounting and mutual settlements of consumed electricity, increases the number of companies and their overhead charges in the sphere of sales, additionally complicates electricity trade.

4.2 Comparison of Models: Criteria, Factors, Competition, and Regulation

The section presents a *qualitative* comparison of the four considered models of electricity market organization. Expediency of transition from Model 1 to Model 2 and the following ones can be substantiated in more detail only for specific countries and based on the analysis of practical experience in operation of different electricity markets.

Economic efficiency and reliability of power supply of the country (economy, social sphere, and population) should apparently be taken as the *key criteria* in comparison of the models. Besides, consideration should also be given to social and environmental impacts of reforms and the national energy security.

The comprehensive analysis in terms of all these criteria is very difficult. Therefore, we will introduce certain simplifications and assumptions. Firstly, suppose that the environmental impacts of restructuring are insignificant, i.e., they are approximately the same in all the market models and this criterion can be neglected. Secondly, as regards the social impacts, we will consider only the changes in the final electricity price for population (its rise or reduction) at transition from one market model to the other. In this case the social criterion is included in the criterion of economic efficiency of power supply.

Furthermore, two of the indicators (factors, indices) of energy security that are associated with scaled and long-run interruption in power supply are (1) large system emergencies (blackouts) and (2) shortage of capacity and electricity that appears in the process of EPS expansion. The first indicator is closely related to the criterion of power supply reliability and the latter will be expanded by adding the probability of system emergencies to it. There is a good reason to separate the second indicator as an independent qualitative criterion: "provision of deficit-free expansion of EPSs."

Hence, comparison of the models will be carried out in terms of three basic criteria:

1. Economic efficiency of power supply
2. Reliability of power supply (including blackouts)
3. Provision of deficit-free expansion of EPSs

Note that an important feature of the first criterion is the *fundamental distinction of the economic efficiency for electricity producers and consumers*. One of the principal objectives of restructuring is to enhance production efficiency owing to competition. Generally speaking, this objective is not necessarily achieved if the competition effect is lower than the expenses on competitive market organization. However, would it even be achieved, it is important, who is the beneficiary—the electricity producer or consumer. If all the effect falls to producer, the economic efficiency of electricity supply for consumers does not increase, i.e., according to this criterion restructuring is ineffective. For consumers the economic effect will be gained only if the final electricity price decreases. Therefore, when the market models are compared in terms of the economic criterion, the following principle should be stuck to "Deepening of reform must yield an effect for both electricity producers and consumers." In other words, it is necessary to reach a compromise of producers' and consumers' interests.

The concepts of power industry restructuring are based on the fact that the competition (that should give an economic effect) can be arranged in the spheres of electricity generation and sale and the spheres of electricity transmission and distribution remain monopolistic and state-regulated. When passing from Model 1 to the following market models, these spheres are successively separated and split into a set of companies. In this context it is important to analyze *how the total costs in the electric power industry are distributed among individual spheres* and what effect of competition can be gained in different spheres.

We will address the conditions, in which UPS of Russia was operating and developing 3–5 years ago without regard to the expenditures on power industry restructuring. We will try to assess the shares of costs in different spheres in electricity tariffs for final consumers (on the average for UPS). Consideration will be given to operating costs in the four spheres (generation, transmission, distribution, and sale) with addition of an investment component of the tariff that is essential for UPS development (construction of new objects including distribution networks). Thus, the operating costs in each sphere are intended only for operation of *existing* objects and their facilities including their upgrading and renewal owing to depreciation charges and also normal profit.

The sphere of electricity transmission covers main grids of UPS (belonging now to the Federal network company "FSK EES") as well as the dispatching control of all levels. The sphere of electricity distribution along with the transmission lines 6–110 kV and somewhere 220 kV also includes low-voltage networks (220–380 V). The number of served transmission lines and substations in the sphere of electricity distribution is, therefore, much larger than that in the transmission sphere. The costs in the spheres of transmission and distribution comprise network losses. The sphere of electricity sale actually embraces only the companies "Energosbyt" of different levels that collect payments from consumers.

In terms of the above explanations Table 4.1 presents the author's expert estimate based on the study of some materials.

Despite possible fallacies of the data in Table 4.1, their analysis allows the following issues to be revealed:

1. The generation costs make up the largest share, and in this sphere the main competition effect may be achieved.
2. Costs in the electricity transmission and distribution spheres total 35%. These spheres remain monopolistic and state-regulated. Hence, during power industry restructuring the costs in these spheres can be reduced only by regulation improvement.
3. The share of costs in the sale sphere is very small. The competition effect here (in the retail markets) will be, therefore, minor in absolute magnitude.

Table 4.1 Estimation of the shares of production costs in different spheres of electric power industry in tariffs for final consumers (on the average for UPS)

Sphere	Type of sphere	Share (%)
Electricity generation	Competitive	55
Electricity transmission	Monopolistic	10
Electricity distribution	Monopolistic	25
Electricity sale	Competitive	5
Power industry development	–	5
Total	–	100

4. The share of investment component in the tariff, which provides the development of the Unified Power System, is not high either. Even if the development gathers pace, the increase in this share will lead to proportional decrease in shares in all other spheres.

The issues of *state regulation and competition effect* are fundamental to compare the market models. The state regulation of electricity tariffs is envisaged in Models 1 and 2. Besides, tariffs should also be regulated in Models 3 and 4 if the shortage arises or conditions for competition are not provided for some time. State regulation is maintained in Models 3 and 4 in the monopoly spheres of electricity transmission and distribution. Hence state regulation is to a larger or smaller degree inevitable in all models of electricity market. For electricity consumers, however, it is beneficial since state regulation prevents market power and unjustified price rise.

Later, consideration will be given to a regulation scheme of tariffs established for electricity producers. This scheme makes it possible to achieve a compromise between the interests of producers and consumers. With the tariffs to be established for a long term (several years), the producers will have a stimulus and time to decrease production costs and gain maximum economic profit. With the next revision of tariffs, part of this profit can be left to producers and the other part can be used to decrease the tariff. This way of regulating tariffs can be applied both in the regulated monopoly and in the single-buyer market model.

Competition Effect A more thorough analysis makes it clear that competition effect does not imply only a direct rivalry among competitors. Indeed, the wish to enter the market and, if possible, to supplant others makes market participants decrease production costs and offer lower prices. However the main driving force of enhancing the production efficiency is the wish of producers to reap the maximum profit. This is a general law of the market economy.

Several types of profit are distinguished in the economic theory [34, 40]: normal, economic, monopoly, and producers' surplus. The profit above normal is considered to be super profit. Here the monopoly profit and producers' surplus do not result from the efficiency enhancement and production cost decrease. They are related to special states (types) of market (which can occur in power industry as well). Only economic profit represents an additional profit above normal, which can be made owing to a successful (above average) activity as a result of introduction of innovations and achievements of technological progress. To gain this profit, the production costs should be decreased (below average costs in the industry), which finally results in the decrease in product price. Therefore, formation of economic profit is encouraged.

In the competition-based Models 3 and 4 in the wholesale electricity market equilibrium, prices will be formed according to demand and supply. With these prices (formed and fixed) the producers in the market will have a stimulus to make the maximum profit by decreasing production cost. This very stimulus will be the second factor (along with the desire to enter the market) providing the enhancement of production efficiency under competition.

Meanwhile, the same wish of the producer to gain the maximum profit can be used for tariff regulation in Models 1 and 2 if the tariffs are established for the period of several years, which is sufficient to real decrease in costs.

4.2.1 Possibilities for Creation of Stimuli to Increase Efficiency of Electricity Production Under Tariff Regulation

With the improvement of state regulation of electricity tariffs (Models 1 and 2), it is important to follow a rather obvious *principle*—the system of tariff regulation should be beneficial for both producers and consumers of electricity. This means that for *producers* there should be a possibility to gain the *maximum profit* by decreasing production cost, yet the tariffs should *decrease* in time (with other things being equal) to the benefit of consumers.

Let us first consider how this can be implemented in the single-buyer market model when competition is among electricity producers. The competition effect in this model manifests itself in two ways:

1. Competition among the *most expensive, marginal* producers for conclusion of a contract with Purchasing Agency (for entry into the electricity market). For them this is essential, since otherwise they will inevitably go bankrupt and cease to exist. For cheap producers (e.g., HPP) participation in the balance is guaranteed, and they do not take part in such a competition.
2. Tendency of *all* producers that have entered into the market to gain the maximum profit if tariffs are *fixed* for rather a long term during which they will be really able to decrease costs.

The first factor of competition should be very effective in decreasing costs of the most expensive producers. However, it does not affect the remaining (the majority of) producers. Therefore it is very important to use the second factor of competition—the tendency of producers to gain the maximum profit. To this effect, electricity tariffs (regulated prices) should be fixed individually for every producer for a rather long time. This is what the single-buyer model suggests—conclusion of long-term contracts with each certain producer (a power plant or a power generation company).

The profit which can be made by the producer above the normal profit, which is traditionally included in the electricity tariffs, will be called *economic* profit. With a fixed price (tariff), the producer can gain the maximum economic profit by decreasing production costs. Should, however, the economic profit be withdrawn from producers under state regulation, they will have no stimulus to enhance the production efficiency.

The compromise between the interests of producers and consumers under regulation of tariffs (in terms of the indicated principle) can be achieved if part of economic profit is left to producers and the other part is used to decrease tariffs.

The contracts concluded by the Purchasing Agency with operating and new producers as mentioned in Sect. 4.1 differ in terms, prices, and conditions for electricity supply. We will consider first a general scheme of establishing tariffs for *operating* producers. It cannot be considered totally novel since some of its points particularly those concerning prolongation of terms for tariff revision have been suggested many times before. For certainty, we will suppose that the contracts with operating producers are made for the period of 3 years (though it can vary).

Based on the points considered above the scheme for tariff regulation can be as follows:

1. Tariff is established for 3 years ahead for each producer (i.e., revised every 3 years).
2. Annual adjustment of tariff is planned to take into account inflation, changes in fuel prices, and other factors that do not depend on producers.
3. Some decrease in the tariff can be envisaged (though not necessarily) by year of this 3-year period (e.g., 1% annually).
4. All the savings of product costs, i.e., economic profit to be made by the producer over 3 years is left to producer.
5. Tariff for the next 3-year period is set on the basis of the tariff for the previous year and actual costs of producer. New tariff is established in a range between the tariff of the previous year and actual costs. Thus, on the one hand, the tariff is decreased (to the benefit of consumers), but, on the other hand, part of economic profit, gained ("earned") by a producer over the previous 3 years will be left to the producer for the next 3 years.

Figure 4.5 shows the tariff and production costs changing during four stages of tariff revision. Here it is supposed that:

- There is no need to adjust tariff, which was said about in point 2 (there is no inflation).
- The established tariff is invariable during 3 years.
- The tariff for the first period is established at the level of the previous tariff T_0.
- For the new period tariff is established strictly in the middle between the previous tariff and actual production costs, i.e., half economic profit achieved by the producer over the previous period is left to the producer for the next period.

Thus, the producer has a stimulus to gain the maximum profit and decrease production costs. Simultaneously, the tariff gradually decreases, i.e., the interests of consumers are pursued. As the tariff of every producer decreases, the averaged tariff of all producers, which influences the prices of electricity sold by Purchasing Agency to consumers, will naturally drop.

The considered general scheme of tariff regulation should certainly be detailed in many aspects. However potentially the scheme makes it possible to create a stimulus for producers to decrease costs similar to the stimuli existing under free competitive market (in Models 3 and 4) and enhance the production efficiency.

Contracts concluded by the Purchasing Agency with new producers (for construction of new power plants) will have their specific features because of longer

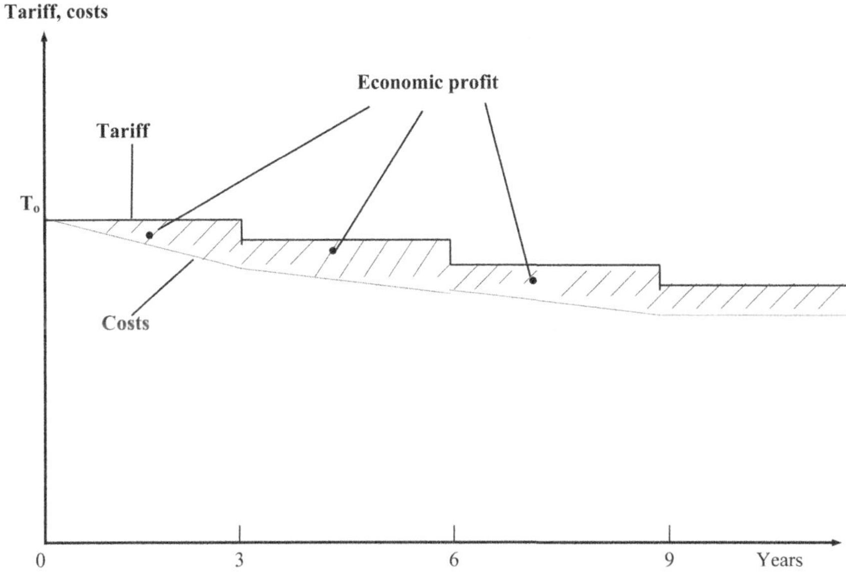

Fig. 4.5 A scheme of tariff regulation

periods (10–15 years), the need to payback investments, etc. Obviously, contracts have to be concluded on the basis of tenders among various companies according to the criterion of minimum electricity price suggested by them. There are no principal difficulties here and we will not dwell on this kind of contracts.

The same scheme can be applied to regulate tariffs in Model 1 to take into account the cost in the generation sphere of monopoly vertically integrated power companies.

4.2.2 Qualitative Analysis and Comparison of Models

Let us start the comparison of models according to the three criteria indicated above (economic efficiency, reliability, deficit-free expansion of EPS) with *Model 1.*

According to the criterion of *economic efficiency* this model is rather good: in vertically integrated companies (VICs), the economies of scale for EPS is implemented to a great extent, whereas with their regulation, the tariffs for consumers are established at the level of average costs of a company. This makes it possible to cover all the costs of the companies, including the investments required for EPS expansion, and provide their normal profit. This model is beneficial for electricity consumers because the tariffs are low.

There are difficulties and, possibly, flaws of the state regulation in this model. However, as is shown above, there are ways to alleviate them, in particular in order to fulfil the ambition of VICs to gain the additional (above normal) profit through a decrease in production cost.

In terms of *power supply reliability*, Model 1 should be considered the best owing to the greatest degree of EPS integrity, responsibility of monopoly companies for reliable power supply, advance and optimal planning of EPS expansion (including provision of capacity reserves), as well as the easiest to implement dispatching control. The reliability level in power industry of the western countries before reforms was 0.9996.

This model also does not have problems with *deficit-free EPS expansion* because the necessary investments are included into the investment component of tariffs. These investments are divided by all electricity produced within the monopoly company, which makes the investment components low (for details see Chap. 6). In some countries, as already mentioned, there was even overinvestment due to the interest of private VICs in the increase in their fixed capital and overcautious behavior of regulatory bodies. This overinvestment, however, can be undoubtedly avoided provided the regulation is more thorough and accurate.

One of the facts that confirms the advantages of Model 1 is that the regulated monopoly VICs are preserved in many countries (France, Japan, most of the USA states, provinces of Canada, etc.). It should be noted, however, that in developing countries with a large territory and population (China, India, and now Russia) the monopoly VICs that cover the entire country become too "bulky" and hard to regulate. Therefore, transition to Model 2 may turn out to be sensible and there can be several purchasing agencies on the territory of the country (e.g., in China).

Analyzing *Model 2* and comparing it by the same criteria we can underline the following points. The *economic efficiency* of Model 2, on the one hand, somewhat decreases due to division of generation sphere. This partly breaks the integrity of EPS and decreases the economies of scale: it will be more difficult to make a potentially possible decrease in capacity reserves; independent generation companies will mainly benefit from implementation of the technological progress achievements in the sphere of generation (i.e., this effect will reach consumers to a smaller degree); the administration and management expenses will rise, etc. However, on the other hand, the economic efficiency will rise due to competition among electricity producers. It is rather difficult (if possible at all) to quantitatively estimate this decrease and rise. For certain countries, the relationship between them may be in favor of either Model 1 or Model 2.

As regards the tariffs for electricity consumers, these will be almost the same as in Model 1 due to similar averaging of tariffs (costs) for producers or even a bit lower if competition effect partly extends to consumers (as in Fig. 4.5). Hence, we can state that according to *the economic criterion* Models 1 and 2 are equivalent and for certain countries, depending on conditions, either model may appear to be the best.

According to *reliability criterion*, Models 1 and 2 should also be almost equal since in both of them it is possible to maintain the required level of capacity reserves (and topology of electric networks) and provide almost equal conditions for dispatching control. A more comprehensive analysis may reveal the factors caused by the division of generation sphere, which can decrease power supply reliability in Model 2. However, this decrease will hardly be large.

In the model of single-buyer market, the *deficit-free expansion of EPS* is provided owing to the long-term (10–15 years) contracts concluded for power supply from new power plants. The Purchasing Agency, which is responsible for power supply, will plan power system expansion in advance similar to the regulated monopoly, i.e., it will optimize the structure and commissioning of generation capacities. The financing mechanism for construction of new power plants will be similar to the financing mechanism for construction of power plants *at the expense of credits* in the regulated monopoly. Payback of investments as well as payback of credits are included in electricity tariffs and are paid by consumers. Some nuances may arise because in regulated monopoly there can also be self-financing (inclusion of investments in tariffs without using bank credits). Self-financing is more profitable than the use of credits at low rates of electricity consumption growth (see Chap. 6). Therefore both models are good according to this criterion, but depending on the state of the economy of a specific country, the financing of EPS generation capacity expansion may turn out to be more efficient under either model of electricity market organization.

Thus, we can state that *Models 1 and 2* have almost the *same qualitative equivalence* by all the three criteria. At the same time for certain countries, either model can turn out to be preferable according to one or another criterion.

It is necessary to remember that both models suggest regulation of electricity prices (tariffs). Shift from either of them to Model 3 or 4 means *deregulation* (or liberalization) of electricity market, i.e., radical changes of the market design. Therefore there can be great differences between the models of deregulated (competitive) markets and considered regulated ones.

Switching to the models of competitive markets we should emphasize *the difference between the economic efficiency* (the first criterion) *for producers and consumers* once again. With electricity price regulation in Models 1 and 2, the compromise between the interests of producers and consumers was provided to some extent (depending on the quality of regulation). After liberation of prices (wholesale prices in Model 3 and retail prices in Model 4) the compromise can hardly be spoken of. The diametrically opposed interests of producers and consumers determine different economic estimation or efficiency of electricity market organization models. *For consumers the economic benefit can be gained only at electricity price decrease.*

The interests of consumers representing all the other sectors of the economy (except for the power industry), population and services, should undoubtedly be of higher priority than the interests of electricity producers. This will also foster the improvement of *social* conditions in the country. Therefore, our assessment of Models 3 and 4 according to the economic criterion will be based on the interests of consumers.

Then, if to think from the viewpoint of consumers, it will be clear that by the *economic criterion Model 3* is obviously worse than Model 2. This is explained by the increase in equilibrium wholesale price in the wholesale market from the level of average costs of electricity production, which was under regulation, to the level of costs of the least efficient marginal power plants in EPS (for detail see Sect. 5.4)—a fact that has been mentioned many times. As was noted in Sect. 4.1, under regulated single-buyer market, only producers compete with one another and this results in decrease in costs in the sphere of generation. Producers have no possibility to use market power and consumers are in the privileged position. Organization of competitive wholesale market in which buyers (consumers) start to compete as well, on the one hand, *in no way fosters* the enhancement of electricity production efficiency (competition of consumers cannot decrease production costs) and on the other hand, consumers (buyers) lose their privileged position while producers getting rid of regulation improve considerably their position in the market and can enjoy market power.

Some countries (e.g., Chile and England) during the first years after transition to the competitive market saw a decrease in the wholesale electricity prices. However, this happened due to a number of some other country-specific factors. Further the influence of this property of the competitive market inevitably led to the price rise.

According to the criterion of *reliability*, Model 3 is behind Model 2 for several reasons:

- Further decrease in EPS integrity with creation of new companies (DSC) with their own interests.
- Complication of dispatching control, particularly in the emergency situations due to the need to take into consideration the supplies under bilateral contracts; arising problems of congestion management; involvement of Trading System Administrator in control of operating conditions, etc.
- Difficulties in maintaining the required capacity reserve as power systems expand (see below the next criterion).

System blackouts that occurred in 2003 in the northeastern states of the USA and the countries of Western Europe (England, Sweden, Denmark, and Italy), which shifted to the competitive market confirm the fact of the reliability deterioration of power supply.

Particular difficulties that arise under the competitive wholesale market concern the investment of construction of new power plants. As was mentioned before (and will be considered in Chap. 6), these occur due to a radical change in the financing mechanism of construction and emergence of the price (economic) barrier for the entry of new electricity producers into the market. These difficulties can lead to (and in many countries have already resulted in) the shortage of generation capacities and electricity. Therefore, in terms of the criterion of *deficit-free EPS expansion* Model 3 is also considerably worse than Model 2.

Thus, in all respects (in terms of all the three criteria) Model 3 is behind Model 2. Hence, it is also behind Model 1, which is almost equivalent to Model 2.

The flaws of competitive wholesale market will manifest themselves in *Model 4* as well. These will be supplemented by the difficulties and costs of the *retail* electricity market organization, which were discussed in Sect. 4.1. In particular, along with administrative expenses, many sales companies emerging in the retail markets will have to bear advertising and marketing expenses that were not borne by distribution sales companies in Model 3. The author did not find any publications comparing quantitatively the potential effect of competition in the retail markets with the costs to be borne for their creation and operation. At the same time there are works that cast doubts on the efficacy of their organization (e.g., [23]). Retail markets, as was already mentioned, were not created in Chile and Brazil. Therefore, we can state that Model 4 is also worse than Model 2 (and Model 1) in all respects.

Hence, deregulation of electricity markets is inappropriate from the viewpoint of consumers. The qualitative analysis made shows that deregulation may lead to negative implications. The next section details them in the context of the experience gained from the operation of competitive markets.

Box 8 Qualitative Comparison of Electricity Market Models

1. The models were compared with respect to three criteria:

   a. Economic efficiency of electricity generation
   b. Power supply reliability (considering blackouts)
   c. Deficit-free EPS development

   According to the first criterion it is necessary to distinguish economic efficiency for producers and for consumers since their interests (in terms of electricity prices) are directly opposite. The models were assessed by this criterion from the standpoint of *consumers*, i.e., increase or decrease in electricity prices.
2. The principal difference of Models 1 and 2 from Models 3 and 4 lies in electricity price regulation in the first two models and its absence in the last two models. If prices are regulated, they are fixed at the level of average (weighted) costs of power companies; if not, the wholesale prices rise to the level of costs of the least effective (marginal) power plants. Thus, Models 3 and 4 are not profitable for consumers.
3. The competition effect leading to a decrease in production costs manifests itself in two ways: first, producers tend to enter the market and remain there, and second, they tend to get maximum profit at the formed market prices. The first incentive concerns the least effective producers, the second is characteristic of all the producers participating in the market.
   The tendency to make the maximum profit can also be used for price regulation, if to impose the producers' tariffs for a long time period (several years). In this case the producers will have both an incentive and time for decreasing costs.
4. The qualitative comparison has revealed that Models 1 and 2 are nearly equivalent in terms of all the three criteria. Depending on the country-specific economic and other conditions either of the two models may be preferable with respect to one or another criterion.
5. Models 3 and 4 turned out worse than the first two models for all the criteria because of rise of the wholesale electricity prices, deterioration of

power supply reliability and problems in investing generation capacities that may cause their shortage. Model 4 is potentially worse than Model 3 due to additional expenditures for organization of retail markets and maintenance of numerous sales companies and also negligible benefit (decrease in costs) that can be achieved owing to competition in the area of electricity sales.

4.3 Flaws of the Competitive Electricity Market

Right from the beginning, there has always been opposition to transition to the competitive market (deregulation) in power industry. A shining example of it is the USA and Canada where most of the states and provinces retain regulated monopoly power companies. Similar opposition exists in Russia. Even at the stage of discussion of the reform conception numerous publications appeared to criticize it [54–57]. The discussion of the conception and its approval "by an order" are described in detail in [58]. The conception has been criticized also after the Law [39] was passed (see, e.g., [20, 21, 58, 59]).

In the past years the implications and progress of reforms in different countries have been actively discussed because the problems and negative consequences have become apparent [2–12]. It is stated that the reforms very often lead to a rise of electricity price, lack of investments, power shortage, deterioration of power supply reliability, etc. As a result the original conceptions of reforms are revised (reforms are reformed), the reform process drags on. The experience of various countries in reforming is considered in detail in Chap. 7. In the current chapter we will just enumerate the electricity deregulation flaws and comment on them briefly.

A profound analysis of deregulation experience was made in [9]. The authors on the basis of an extensive review of 114 publications found 11 difficulties, flaws and negative consequences in organization of the competitive electricity markets. Many of them were also pointed out in other publications indicated above. Not listing all the 11 we will indicate only the main drawbacks from our viewpoint:

1. Considerable costs of competitive market organization and operation
2. Market power in the sphere of generation
3. Extreme volatility (and unpredictability) of wholesale market prices
4. Lack of investments in expansion of generation capacities and electric networks
5. Deterioration of power supply reliability
6. Electricity price rise (in many markets)
7. The problem of compensation for stranded costs
8. Deregulation benefit, if any, is gained mainly by producers (not consumers)

Let us explain in a nutshell the problem of stranded costs. It is associated with the fact that prior to restructuring vertically integrated monopoly companies had unpaid credits taken to construct power plants and transmission lines as well as other financial obligations that were supposed to be paid back at the expense of tariffs. With the split of monopoly companies, on the one hand, it was not always clear what company had to payback the debts and, on the other hand, under the competitive market an individual company was not always capable of paying back the debt ascribed to it from its own revenues. The problem of compensation for stranded costs in different countries was solved differently. Great Britain and some states of the USA, e.g., allowed the tariffs to be increased several years before the restructuring. This increase was one of the reasons for electricity price decrease in some countries (states) during the first years after creation of the competitive market.

The authors of [9] illustrate each of the identified flaws by the examples of definite markets in the USA, Canada, Western Europe, Australia, South America, and other countries. They present, in particular, the figures of about $1 billion for market organization and about $100–250 million yearly for their operation.

It should be emphasized that the flaws were revealed on the basis of *real experience* of organization and operation of the competitive electricity markets. The authors just state and generalize the facts not aiming to explain the revealed disadvantages. Their article is related to the reforms planned in the power industry of Israel and the main conclusion made is that it should be done very cautiously.

Meanwhile almost all the indicated drawbacks of the competitive markets (Models 3 and 4) have already been discussed earlier in the chapter while considering and comparing the models of electricity market organization. They are *theoretically proved* and an attempt to substantiate them in the book is made by the author with the use of the works by other researchers known to him.

The main cause of these drawbacks is *imperfection* of electricity market and specific features of EPS and electricity as a commodity. *Liberation of* electricity *prices with no conditions for perfect competition* should knowingly lead to negative consequences.

Professor F.E. Banks, an expert in the field of economics and finances touches upon several points in [5, 65]:

- Transformation of the natural monopoly into some textbook example of perfect competition is almost impossible and will make no sense under any circumstances.
- Deregulation of electricity market raises uncertainty (both for producers and consumers) which leads to:
  - A sharp decrease in investments in new capacities, even in combined cycle plants as was observed in Brazil.
  - Extraordinary volatility of prices in the spot markets.
  - The absence of consumer wish to change suppliers (in retail markets), i.e., to use a widely advertised "possibility of choice."

- In the competitive markets pricing based on the short-run marginal costs becomes a norm. This may essentially raise the prices since all the generated electricity will be sold at a price corresponding to the marginal costs of the marginal generator.
- Under market deregulation the regulated monopoly turns into oligopoly. The number of oligopolists can be decreased through merging or mutual acquisition of companies, which makes the oligopoly stronger. As soon as the possibility appears to use market power, the oligopoly will do that to raise electricity prices. Even if in some countries and regions the prices saw a decrease they are sure to climb again.
- After creation of the market of the Scandinavian countries (Nordel) electricity prices in Sweden and Norway rose additionally because of electricity export to other countries including Germany.
- Due to clear distinctions between electricity market and other markets, the risks associated with uncertainty of electricity prices cannot be insured in the traditional financial markets of derivatives (futures, options), which work perfectly well for other goods and securities.
- Since power industry deregulation fails even in such a wealthy, technically developed, and market-oriented country as the United States, it will not be implemented anywhere in the world at least not in the long run.

The analysis made by Professor F.E. Banks additionally illustrates the nature of the competitive market flaws.Two more disadvantages should be added to the above list. These will also be discussed in the book:

9. Difficulties (or even impossibility of) in substantiating the construction of intersystem (including interstate) electric ties that realize the capacity effect of interconnecting power systems, which was considered in Sect. 2.2. This is explained by separation of electricity generation from transportation and creation of independent companies: the construction costs of intersystem ties should be borne by network companies, while the benefit (the generation capacity decrease) will be gained by generation companies.
10. Electricity export is unprofitable for consumers of exporting county since it raises the demand for and prices of electricity; the benefit in the exporting county will be gained only by electricity producers, moreover the benefit will be double—from export and from rise of price for all electricity consumed in the country. Meanwhile under price regulation export can be (and was) profitable for consumers of both countries—internal prices in the exporting country decrease at the expense of export revenues.

In Box 9, the disadvantages of competitive electricity market are put in a little different wording and order. In Chaps. 5 and 6 they will be analyzed and substantiated in more detail, with quantitative examples. Chapter 7 presents their illustration by the practical experience of competitive markets operation in various countries of the world.

Box 9 Basic Drawbacks of Competitive Electricity Markets (Models 3 and 4)

1. Considerable costs for organization (creation) and operation of competitive markets that amount to several hundred million dollars.
2. Increase in the wholesale electricity prices from the level of average costs for EPS as a whole (at price regulation) to the level of costs of the least economically efficient (marginal) power plant.
3. Extraordinary volatility (and unpredictability) of prices in the spot electricity markets.
4. Problems in investing in generation capacity expansion due to emergence of the price barrier for new electricity producers.
5. Freedom of electricity producers from regulation and creation of conditions for them to form oligopoly and use market power by manipulating the prices or forming power shortages, in particular, by ceasing to build new power plants.
6. Decrease in power supply reliability.
7. Challenges in substantiation of constructing intersystem electric ties that realize a *capacity* effect of EPS interconnection.
8. Electricity export ceases to be mutually advantageous.
9. The deregulation effect, if any, is obtained mainly by electricity producers, not consumers.

The indicated drawbacks are revealed by the theoretical analysis and proved by practical experience of operation of competitive electricity markets. In the following chapters they will be substantiated in greater detail.

Chapter 5
Short-Run Production Costs and Electricity Markets

This chapter is dedicated to the specific features of cost characteristics of various types of power plants and power generation companies (PGCs) in the short run (with their installed capacities being fixed). Consideration is also given to price formation in the wholesale electricity market under different models of its organization. This chapter starts with the analysis of *differences* between short-run and hourly electricity production costs (Sect. 5.1) and shows the general inefficiency of spot markets (Sect. 5.2) which are usually organized during the transition to competitive electricity markets (Models 3 and 4). Then the characteristics of short-run (annual) costs of *individual power plants* are studied on a quantitative example of European section of Russia's Unified Power System (Sect. 5.3). In Sect. 5.4, the same example is used to analyze the costs of *generation companies* and the *wholesale price formation* under different models of electricity market.

5.1 Relationship Between Short-Run (Yearly) and Hourly (Instantaneous) Costs of Power Plants and EPS Generation Sphere

It is very important for further analysis to clearly distinguish between *short-run* costs (in terms of microeconomics) that underlie the *formation of prices* in the wholesale electricity market and *instant* (let us consider them *hourly*) costs which are used for optimization of electric power system (EPS) operation. These notions are often confused, which leads to erroneous conclusions and results.

The power industry is apparently the only (unique) industry where this distinction should be made because of consumption load *variability* within hours of a day, days of a week, and seasons of a year as well because of *inseparability of the* electricity production and consumption processes. Two special features of EPSs considered in Sect. 2.3 are important here:

1. *Mutual dependence* of electricity production *processes* of all power plants that belong to an EPS. Power plants unlike companies in other branches do not enter

DOI 10.1007/978-1-4419-5612-5_5, © Springer Science+Business Media, LLC 2011

the market with "ready-made" and independently produced commodities. Electricity at each time instant is produced *jointly* by all power plants under *centralized* control and *optimization* of operation conditions in the entire EPS.
2. *Optimization* of power plants *operation* with respect to their *instant* (*hourly*) characteristics while the total costs of power plants are determined by the integral results of their operation throughout *the whole year*. This property of an EPS is particularly often underestimated and misunderstood, which leads to erroneous approaches in the economic theory of power industry, in particular to the organization of spot electricity markets.

5.1.1 Mathematical Expression of Links Between Short-Run and Hourly Costs

Let us introduce first *the denotations* for the basic values.

For the *short* run (a year) we will assume generally accepted notations similar to those used in Sect. 3.1 (omitting only the letter "S" for the short run):

Q_i—power production at the *i*th power plant in the short run (a year), kWh/year;
Q_{EPS}—the same for the entire EPS, kWh/year;
FC_i, VC_i, TC_i—short-run fixed, variable, and total costs of the *i*th power plant (for the entire period), \$/year;
VC_{EPS}—short-run (annual) variable costs in the generation sphere of the entire EPS, \$/year;
AFC_i, AVC_i, ATC_i—short-run *average* (specific) fixed, variable, and total costs of the *i*th power plant, \$/kWh;
MC_i—short-run *marginal* costs of the *i*th power plant, \$/kWh.

For the indicated values, expressions (3.1) and (3.2), presented in Sect. 3.1, hold true. Short-run *average* costs AFC_i, AVC_i, and ATC_i are determined by dividing annual costs FC_i, VC_i, and TC_i by the annual power production Q_i. These short-run costs in the generation sphere of EPS will be analyzed in Sects. 5.3 and 5.4. They are taken into account in price formation in the wholesale electricity market (differently depending on market organization model).

To denote *hourly* costs we will apply an additional letter "*H*" and an index "*t*" that means a certain hour *t*. Hourly costs used for the optimization of power system operation are *variable* costs in themselves. These are mainly fuel costs related to the fuel consumption for electricity production. They do not contain a fixed component, which the short-run costs contain. This is correct in terms of both the economics and the methodology—the fixed costs of power plants *should not be taken into account* during the optimization of hourly (instant), daily, and even seasonal power system operation conditions.

In the context of these explanations, we will introduce the following denotation of *hourly* values:

N_{it}—mean hourly power (load) of the ith power plant at hour t, kW
HVC_{it}—hourly variable costs of the ith power plant at hour t, \$/h; they are the function of (depend on) power N_{it}: $HVC_{it} = f(N_{it})$ or $HVC_{it}(N_{it})$
HVC_{EPSt}—hourly variable costs of EPS generation sphere at hour t, \$/h
$HAVC_{it}$—hourly average (specific) variable costs of the ith power plant at hour t, \$/kWh
HMC_{it}—hourly marginal costs of the i-th power plant at hour t, \$/kWh

All these values have the meaning of "hourly average," since the consumer load and operating powers of power plants vary during an hour. The "hourly" characteristics are widely used in the optimization of power system operation, though an hour can be divided into shorter intervals. Hourly intervals are convenient because the mean hourly power N_{it} (in kilowatt) simultaneously equals the electricity (in kilowatt hours) produced during this hour.

Consider now some *relationships* between short-run (yearly) and hourly values.

First we will write the following, quite obvious, relationships:

$$Q_i = \sum_{t=1}^{8760} N_{it}, \tag{5.1}$$

$$Q_{EPS} = \sum_{i=1}^{I} Q_i = \sum_{t=1}^{8760} \sum_{i=1}^{I} N_{it}, \tag{5.2}$$

$$VC_i = \sum_{t=1}^{8760} HVC_{it}(N_{it}), \tag{5.3}$$

where I is the total number of power plants in EPS.

In these relationships, the operating powers of power plants N_{it} are not set or taken arbitrarily. They are chosen and assigned in the process of optimization of power system operation. The optimization is carried out according to the criterion of minimum short-run variable costs of the entire EPS throughout the entire period at issue (a year):

$$VC_{EPS} = \min_{N_{it}} \sum_{t=1}^{8760} \sum_{i=1}^{I} HVC_{it}(N_{it}) \tag{5.4}$$

with numerous additionally imposed constraints.

Let us not delve into the optimization process. It is rather complex. Cyclic changes in loads, water inflow to hydropower plants, thermal load at cogeneration power plants, and other factors may call for successive optimization of first seasonal, then weekly, daily, and hourly EPS operation conditions. Normally, the load of a power plant N_{it} at some hour t is set according to the results of optimization of *daily* EPS operation conditions (taking into account fuel consumption for starts–stops of TPP units, limited daily volumes of water at HPP, etc.). However, if EPS operation can be optimized for a *certain* hour t, the optimality criterion will be the minimum hourly variable costs throughout the entire EPS:

$$HVC_{\mathrm{EPS}_t} = \min_{N_{it}} \sum_{i=1}^{I} HVC_{it}(N_{it}). \tag{5.5}$$

Here, hourly marginal costs of power plants should be used:

$$HMC_{it} = \frac{dHVC_{it}}{dN_{it}}. \tag{5.6}$$

Optimization of EPS operation for all days and hours of a year will determine the optimal mean hourly operating powers of power plants,

$$\bar{N}_{it,} \; i = 1, \ldots, I, \; t = 1, \ldots, 8760, \tag{5.7}$$

which, in turn, will be used to determine annual electricity production Q_i (5.1) and annual variable costs of power plants VC_i (5.3). In principle, this can be learnt only at the end of the year by actual results of EPS operation. The results will vary from year to year. At the beginning of the forthcoming year we can only make forecasts or preliminary plans for these values. This adds considerable uncertainty to the determination of short-run (yearly) costs of power plants. It, in particular, creates difficulties in establishing electricity tariffs under state regulation of prices (in Models 1 and 2).

Expressions (5.3)–(5.6) employ only *variable* costs of power plants and the generation sphere of EPS. Meanwhile, the *real value of electricity* and *its price* are determined by the total costs including fixed costs as well. As applied to the hourly intervals, as was already mentioned, the total costs make no sense at all. Therefore, it is necessary to consider *short-run total* costs (TC_i and ATC_i) throughout the entire period (a year). *Average* (specific) total costs are the most important:

$$ATC_i = FC_i/Q_i + VC_i/Q_i, \tag{5.8}$$

where fixed costs FC_i can be considered known (set, fixed) and variable costs VC_i and annual electricity production Q_i are determined by expressions (5.1) and (5.3). ATC_i characterizes the real value of 1 kWh of electricity produced by the ith power plant and should be used to determine electricity price.

5.1.1.1 Analysis and Discussion of Interrelations

As is seen from the above written relationships, *hourly* (instantaneous, mean hourly) costs of power plants are used (and with good ground) for optimization only. Such optimization is necessary due to cyclic variations of load, changes in the mix of generation equipment, electric network scheme, etc. Based on the optimization results the operating powers (loads) N_{it} are assigned to each power plant at one or another hour of a year.

The values of N_{it} are mostly *random* (or uncertain) since they depend on many factors: consumer load at hour t, which varies randomly, the mix of generation capacities by type of power plants, actual state of power plant operating equipment with respect to its emergency rate and the need for maintenance, filling of reservoirs of hydropower plants in power system, etc. The value N_{it} can be forecasted with a lead time of several months in a probabilistic (ambiguous) form only.

At the same time, the sum of values N_{it} for all 8,760 h of a year determines the annual electricity output Q_i by the given ith power plant. Hence, the value of Q_i will also be to a greater extent *random*. Particularly, this applies to hydropower, wind power, and solar power plants whose annual output depends on random natural factors.

The random character of the values N_{it} and Q_i creates *uncertainty of short-run* (annual) costs of power plants. The values of N_{it} affect the hourly costs HVC_{it}, and the sum of the latter determines *annual variable* costs VC_i (5.3). Simultaneously, the annual production Q_i influences *short-run average total* costs ATC_i (5.8). Thus, the values VC_i and Q_i will largely be random or uncertain in the expression (5.8). Accordingly, there will be uncertainty about the short-run average variable costs AVC_i, and about the short-run marginal costs MC_i that depend on them.

The *uncertainty of the short-run power plant costs* should be considered as a *specific feature of the power industry* which makes it distinct from the other industries. This feature is conditioned by such EPS properties as load variability within hours of a day and seasons of a year, the inseparability of electricity production and consumption processes, and joint electricity production by all power plants within the EPS at each time instant under centralized control of this production. There are no such properties and specific features in the other industries.

As was already mentioned, the uncertainty of short-run costs in the generation sphere creates *difficulties in state regulation* of electricity tariffs *for electricity producers*. These difficulties are less considerable with the tariffs established for the whole vertically integrated company (VIC) (Model 1) when the total EPS electricity output Q_{EPS} (5.2) and the total variable costs of the EPS generation sphere VC_{EPS} (5.4) are only important. In this case the uncertainty is largely conditioned by random changes in the total electricity consumption in EPS and annual water inflow to HPP. It is more difficult to impose tariffs for individual PGCs when market is organized according to Model 2. However, these difficulties must not be considered insuperable. The appropriate adjustments of tariffs during a year can be envisaged or the last year's deviations can be taken into account while establishing tariffs for the next year.

The uncertainty of short-run costs will manifest itself to much greater extent in the *competitive* market (Models 3 and 4). It will affect the producer's behavior when the bids are submitted to the spot markets or the long-term bilateral contracts are concluded with consumers (buyers). This will result in an increase in variability and a general price rise in the spot markets and overpricing in the long-term contracts.

The main conclusion to be made from the analysis of relationships (5.1)–(5.7) is that the *hourly costs of power plants consist of variable costs only*. They do not show the *real value of electricity* and, therefore, cannot be used for establishing its prices. The attempts to organize real-time electricity trading (in the spot markets) *contradict the fundamental principles of the economic theory* (microeconomics). Electricity should be traded on the basis of *short-run total costs* (or, rather, long-run costs).

Organization of spot electricity market is incompatible with the theory of microeconomics, which should be considered as another important feature of the power industry that makes it distinct from other industries. Spot markets are considered in more detail in the next section.

Box 10 Distinctions in Hourly and Short-Run Costs of Electricity Generation

1. The hourly (instantaneous, mean hourly) costs of power plants are used to *optimize EPS operation conditions* at cyclical changes of consumer loads. They include solely *variable costs* of electricity production and are not representative of its *real* value. Hence, they *cannot be used for electricity price quotation*.
2. The short-run costs are determined for the whole *short-term period* (*a year*). Both the *variable and fixed* costs of electricity generation are considered in this case. The short-run *average total costs* (*ATC*) are the most significant quantity. These costs reflect a real value of 1 kWh of electricity and must be applied for quotation (formation) of the wholesale electricity prices.
3. Distinction between the short-run and hourly production costs (to be more precise, the necessity to use the latter) is a *unique feature of the electric power industry that depends primarily on such EPS properties as continuous consumption changes, inseparability* of electricity production and consumption processes, and the necessity of *centralized optimization* of joint electricity production by all power plants in EPS at every moment.
4. The short-run costs of power plants and the EPS generation sphere as a whole are characterized by high *uncertainty*. It is also a *specific feature* of the industry that causes *problems* under *state regulation* of electricity prices and *complicates* the process of price formation in the wholesale *competitive* market of electricity.
5. It is impossible (theoretically incompatible) to *organize trade in real time*, because the actual electricity value (and price) cannot be determined at some instant (an hour) of time. This can be done just for the whole *short-term period* (a year) on the basis of the *short-run total* costs of electricity production. Impossibility to organize spot trade can be considered as an additional feature of the power industry that distinguishes it from other ones.

5.2 Spot Electricity Markets: Pitfalls in Their Organization

The short-run costs and electricity markets should be analyzed starting with the spot markets that, as was noted in Chap. 4, are not actual short-run markets with respect to the electric power industry in terms of microeconomics. They cannot provide an adequate (correct, perfect) electricity trade, and therefore they must be simply excluded from consideration in case of short-run electricity production costs and markets. Yet, they have gained such a wide spread (including the NOREM conception in Russia) that their unacceptability should be shown specially.

In general sense, a spot market means that a seller delivers goods *immediately* and a purchaser pays for them “on the spot” [38]. This is indeed the case with trade in the markets of many kinds of goods, starting from vegetables and fruits. The current markets of such goods as oil or coffee seem to be very complicated because of the large volumes of trade and computerization. The basic principle of trade is, however, the same—immediate delivery of and payment for goods.

The spot markets are characterized by the high price volatility that is caused by the formed conjuncture of supply and demand as well as the forecasts of future conditions. A seller or purchaser of most goods may prefer to wait for an opportunity to sell or purchase later to advantage. Insurance against risks caused by the price volatility is performed through organization of the so-called secondary markets or the markets of “derivatives”—futures, options, etc.—that will not be considered here in detail. It should only be noted that somewhere the markets of “derivatives” are applied in the electric power industry in addition to the spot markets (such markets are planned in Russia as well). At the same time in [5, 60], the markets of “derivatives” in the power industry are shown to have no prospects.

By analogy with the markets of other goods the spot markets are foreseen in the competitive electricity markets of many countries. In the power industry, it implies electricity trade in real time—with hourly and half-hourly intervals (there were also suggestions on even shorter intervals). We will dwell on “day ahead” markets, though balancing markets also belong to the spot ones.

In the original conceptions of the competitive electricity market, the spot markets played a very important role [61, 62]. They were to provide electricity sale at prices corresponding to actual costs for its production based on variability of consumer demand by hour of the day, day of the week, and season of the year, on the one hand. Electricity prices were supposed to vary by hour of each day (and from season to season) and be quoted by the marginal costs of the least effective power plants.

On the other hand, the spot markets must (as was supposed) give “price signals” for market expansion (or shrinkage), as is the case for the markets of other goods. In the context that at that time (and nowadays) electricity consumption has been increasing in all countries; in fact, consideration is given to the attraction of investments in expansion of generation capacities of EPS. Hence, the spot markets were expected to give signals providing EPS expansion and no other measures were planned, i.e., construction of new power plants was left to the “invisible hand” of the market.

The tendency to organize such electricity markets may be treated as natural. Similar to the markets of other goods, they promised some advantages. Firstly, they were to ensure trade at "fair" prices reflecting changes in producer costs with changes in consumer load, i.e., rise of prices during the hours of maximum loads and their decrease at the periods of minimum loads. This would force consumers to shift electricity consumption from the hours of maximum load to those of minimum load, which would lead to leveling of load curves and decrease in the absolute value of maximum loads of EPS. Hence, the total required installed capacity of power plants would fall and their more uniform operation conditions would be achieved.

Secondly, the state regulation of electricity prices that causes the known difficulties and is not always effective would become unnecessary. In particular, this refers to regulation of EPS generation capacity expansion that often gave rise to "overinvestment."

The other positive aspects of such markets were anticipated as well. They are, for example, transfer of "risks" of uncertainty of future conditions from consumers to producers. As a result, the latter have to make more correct decisions, "easing thus consumers' lot."

The developers of the concepts of spot markets proceeded seemingly from rather obvious principles:

- It will be possible to create conditions for *perfect* competition in the spot markets.
- The curves of power plant costs are *like* the characteristics of "typical" firms that are considered in microeconomics. In particular, the producer supply curves coincide with the marginal costs and cover total short-run costs.
- Producers will enter the market with the offers that correspond to the marginal production costs and consumers will respond to the formed prices etc.

Let us briefly comment upon these principles. The general *imperfection* of electricity market (regardless of the models of its organization) has already been discussed in Sect. 3.2. It very likely concerns the spot electricity market to the greatest extent. For example, how can new producers freely enter during an hour in the market with the formed deficit and price rise? Or, how can electricity consumers *keep track of prices* in the spot market *in real time* and respond to their changes? It might be feasible only for individual large consumers. Even the sales companies that can track the spot prices are unable to "instantly" inform all their consumers about them.

As shown in the previous section, the main thing, however, is that the *hourly* costs of power plants are by no means *short-run* costs (in the microeconomics sense) underlying the price formation in the "normal" markets. Neglecting this fact should be recognized as an obvious mistake of spot market developers. Seemingly, the traditional optimization of EPS conditions on the basis of hourly costs has misled them.

Finally, the assumption on the entry of producers in the spot market with the offer corresponding to the hourly marginal costs seems rather unrealistic. On the one hand, a producer is free to choose his price offers and can, in principle, ask for any price. On the other hand, as will be shown below, there are reasons caused by specific features of the spot electricity market that just "force" him to deviate from the hourly marginal costs in the offered prices.

The spot markets were attached such a great significance that in Great Britain and California, being among the pioneers in organization of competitive markets, at first the bilateral contracts between consumers (purchasers) and producers were even forbidden. All the trade in electricity was performed through the spot market (the "day ahead" market).

Meanwhile, experience in operation of the spot markets has revealed their grave drawbacks which are described extensively in [5, 9, 63, 64].

The *first drawback* is the extreme volatility and unpredictability of prices. The prices change from zero (and even negative values in some markets) to the values that are several times and even tens of times higher than the actual production costs of power plants up to the price cap, when established. Thus, the hopes that the prices will reflect production costs have obviously not been justified. The reasons are: application of the so-called price-taking offers, in which producers indicate only electricity (or capacity) supplied without its price (if such offers fully cover consumer demand, then the zero prices are established in the market); unfair strategic behavior of producers; weak response of consumers to price changes, etc.

The *second drawback* of the spot market is considered in Sect. 5.1—participation in it of producers only with their *variable* (*hourly*) *costs* (that determine marginal costs). The fixed costs of power plants in this case are not compensated (paid back) (the next section will dwell on this in greater detail). In this context the spot electricity market has to be supplemented by the payment for capacity and markets of ancillary services (maintenance of reserves, frequency, voltage, etc.). As a result, the electricity trade becomes highly complicated, producers are able to manipulate with price offers, and finally the spot market prices change sizably and do not correspond to the actual electricity production costs.

It should be emphasized that the concepts of competitive wholesale markets in different countries have certain distinctions. In Great Britain, for example, there is no "day ahead" market at present, but a balancing market only. In other countries there are two main varieties of spot markets:

1. The "day ahead" market without payment for capacity (*the energy-only market*), for example, in Australia and Scandinavian countries.
2. The "day ahead" market supplemented by the capacity payment or by the short-run capacity market (in some US states and some European countries).

Usually, in addition to these markets, the electricity (capacity) is also traded by the bilateral contracts between producers and purchasers (sales companies or directly consumers) and in the balancing market. Sometimes the mentioned markets of ancillary services and the markets of "derivatives" are organized. Below we will focus on the "day ahead" market without dealing with other types of markets.

Let us consider the situations in which the electricity producers are found when submitting offers in the spot market. Naturally, the key incentives of their behavior (objectives, criteria) will be, first, the tendency to penetrate into the market (for the offer to be accepted) and, second, gaining the maximum profit. The first incentive is particularly important for the most expensive producers (with high costs) that will be marginal in the considered hour of the day and may turn out to be "outboard." Potentialities for "penetration" into the market will depend on the general relation between supply and demand in EPS (surplus or shortage of generation capacities) and on consumer loads changing during a day. The profit will be determined by the equilibrium price formed by the offer of the marginal producer and the own expenditures of the producer during the corresponding hour of the day.

Submitting offers, the producer is in rather *uncertain* conditions:

- The total load of EPS consumers during one or other hour of the forthcoming day is not known completely, though for a day ahead the load is forecasted rather well.
- The offers of other electricity producers (competitors) and the bids of purchasers are not known.
- The main thing is that it is unknown whose offer will prove to be marginal and what the equilibrium prices of the spot market by hour of the day will be.

Thus, the trade through the spot market is factually a *game* and a highly complicated one. Since the producer is free to choose prices in the offers, he must maximize his profit, being under uncertainty. He will choose prices based on the available information on the market state, his intuition, inclination to risk and possibly some calculations. The key players here become the most expensive producers, who run a risk to be not attracted during some (or even all) hours of the day, on the one hand, and determine an equilibrium price of the whole market by their offers, on the other hand. They will hold up prices in every possible way during the hours, when they have a guarantee for participation in EPS balance to compensate for outages during the hours (and days) with low consumer loads. Certainly, the rest (more effective) of producers are also interested in such rise of prices that increases their profits.

After "playing" in the market for some time, each producer will elaborate his best strategy of submitting price offers. This can be done most easily by power-generating companies (PGCs) that own different types of power plants. Such PGCs can "play" with offers of their most expensive (marginal, peak) power plants to gain the largest profit from the most effective (base) power plants. The producer strategy concerning offers to the "day ahead" market may be combined with his participation in the balancing market and also in the markets of capacity, ancillary services, and "derivatives," if the latter are available. These strategies will differ depending on whether the spot market is supplemented by the payment for capacity.

In the spot markets *without payment for capacity* producers cannot submit price offers in accordance with hourly variable costs, since their fixed costs in this case will not be repaid. Therefore, they have to hold up offered prices versus the hourly marginal costs. If the power plant loads are set by the Trading System Administrator based on such price offers of producers, the power system operation will not be optimal. Hence, in such spot markets there will be not only high volatility of prices, but also the increase in factual costs of EPS as a whole caused by nonoptimality of its operating conditions.

In the spot markets combining *the payment for capacity* that compensates for fixed costs, the price offers will be, most probably, close to the hourly marginal costs of power plants. However, producers will derive greater possibilities (freedom) for choosing a strategy of submitting offers according to the above-mentioned criteria. The experience shows that this will finally lead to the overall rise of electricity prices. Specifically, it was one of the reasons for cancellation from the "day ahead" market in Great Britain.

The *third drawback* of the spot market is insufficient stability of its "price signals" to attract investments in generation capacity expansion and, *above all, their very late arrival*, when the power shortage has already been formed. The part of high spot prices that exceeds production costs is often called an "investment premium" [65]. This premium was supposed to be used for the construction of new power plants. Meanwhile, there are at least two circumstances complicating or impeding such construction.

Firstly, the premium is received only by *operating* producers that already participate in the market. New producers may be attracted only by a *prospect to get it* in the future. The operating producers getting an "investment premium" will not necessarily use it for the construction of new power plants. This is beyond their interests, since additional capacities will increase the supply in the market and, hence, decrease prices. Producers' interests were analyzed in Sect. 3.2 concerning oligopoly.

Secondly, in order to attract investments of both operating and new producers, the "investment premium" should be high enough, on the one hand, and rather long (10–15 years), on the other hand, for the investments to be paid back. The mentioned unpredictability and volatility of prices in the spot market, however, give rise to doubt of the potential investor about the investment payback. The appearing power shortage leads to high prices at first only during hours (and seasons) of maximum consumer loads in EPS. "The investment premium" in this case will be relatively small. Only grave deficit and persistently high prices in the spot market will make it sufficient for the investor. However, he should be sure that high prices be guaranteed for 10–15 years required for the construction of a power plant and investment payback. But such firm confidence is impossible, since in the competitive market there are several independent (and competing with one another) PGCs and, besides, new producers can enter the market. Each producer independently plans expansion of his capacities and makes decisions on investments. In doing so, each of them runs a risk that the other producers would also start constructing new power plants. Then, the deficit will disappear, prices will go down, and investments will not be paid back (be "lost" partially or fully). Hence, the spot market cannot, in principle, guarantee timely and the more so optimal expansion of generation capacities of EPS. This needs consideration of the electricity market in the long run (see Chap. 6).

Besides the described three drawbacks there are *physical barriers* for new producers in the short run; *inconsistency of the shape of curves for short-run costs of power plants* with the U-shaped type of costs of "typical firms" (see the next section); emergence of the *price barrier* in the competitive market for NPPs in the long run (see Chap. 6), etc. Hence, the revealed drawbacks and general unsoundness of the spot electricity markets may be considered quite logical. This is confirmed by the refusal of England, Brazil, and some other countries from spot markets (with transition to electricity trade by the long-term contracts).

At the same time some countries preserve spot markets and continue to "struggle" with their drawbacks, sophisticating thus even more the trade in favor of electricity producers, as will be shown later.

Box 11 Spot Electricity Markets, Their Drawbacks and Unsoundness

1. The spot markets were organized for trade in electricity in real time (with the hourly or half-hourly intervals). They were to:
   a. Ensure trade at "fair" prices reflecting changes in producer costs by hours of a day (weeks, seasons);
   b. Give "price signals" for the development of EPS generation capacities.
2. Experience in operation of the spot markets has revealed their serious drawbacks:
   a. The spot prices are extremely volatile (and unpredictable);
   b. The fixed costs of producers are not compensated, and therefore the spot markets should be supplemented by payment for capacity, markets of ancillary services, etc.;
   c. The "price signals" of spot markets are unstable to attract investments and received *too late*, when the capacity deficit has already been formed (and its elimination requires several years).
3. Drawbacks of the spot electricity markets are explained by somewhat unrealistic principles underlying their conceptions that do not take into consideration many properties and features of EPS. The actual costs of electricity generation and hence its value and prices can be determined only for the whole short-term period (a year). The attempts to set prices for each hour of the day are simply unfit.
4. The spot electricity markets are not real short-run markets and will not be considered in the subsequent analysis of short-run costs of electricity production and short-run markets.

5.3 Short-Run Costs of Power Plants

This section analyzes the curves of costs for different types of power plants. Consideration is given to the *short-run* costs of operating power plants (at the fixed installed capacities). The shape of curves of power plant costs is compared with the "typical" curves of the firms presented in Sect. 3.1.

5.3.1 General Conditions, Dependences, and Assumptions

To be more precise and certain we will proceed from the following assumptions:

1. The short-run period, during which the installed capacities of power plants remain invariable, is equal to 1 year. Correspondingly, the full costs of power plants (*TC*, *FC*, and *VC*) are determined as the annual costs (for the whole year), and the curves for average costs (*ATC*, *AFC*, and *AVC*) are plotted as a function of their annual electricity generation.
2. Each power plant will be represented so far as an individual firm. Their integration in PGCs or VICs will be dealt with below.
3. The costs of only large power plants that are directly connected to EPS and participate in its territorial market will be analyzed. Specifically they are: hydro (HPPs), nuclear (NPPs), condensing (CPPs) on coal and natural gas, cogeneration (CGPPs) power plants on coal and natural gas. Two types of CPPs on gas will be considered: with conventional steam turbine installations (STIs) and with combined-cycle installations (CCIs).

Power plants of diverse types differ in the mix of costs (fixed and variable), operating conditions, etc. For example, the variable costs of HPPs and the other power plants on renewable energy sources (RESs) are virtually equal to zero. Besides, the annual electricity production by HPPs that determines the average total costs (*ATC*) depends on hydrological conditions, i.e., it is a random variable. If possible, NPPs participate in meeting base loads, while the other power plants cover semi-peak and peak zones of the load curves. The costs of electricity production by CGPPs depend on the volume of heat energy production (on outdoor temperatures that are also subject to random changes). Of great importance for CGPPs is the method of dividing costs between heat and electricity production. There are many other specific features and distinctions in power plant types that will be addressed later, when needed.

It is advisable to analyze the costs of power plants on the basis of their quantitative relationships. Then along with the *shape* of the curves for average costs for different types of power plants, it is possible to compare their *numerical values*, which is important to construct the aggregate supply curve of electricity producers for EPS and the market as a whole. *The European part of Russia's UPS for the conditions* that may be formed roughly at the 2010 level (with regard to fuel prices and structure of power plants) is chosen as a concrete example.

Table 5.1 presents the structure of power plants that is expected in 2010 in the European part of UPS (including IPS of Ural) and their basic indices that are needed for the determination of costs or will be needed later. The figures in the table must be taken as conventional, showing only a general relationship of the indices for different types of power plants. In the columns for CGPPs the values of fuel component in costs for electricity generation remained uncertain, as far as it depends on the volumes and conditions of heat energy production and on the methods of sharing the total costs at combined production between these two types of energy. The characteristics for CGPP costs will be considered in greater detail later.

Table 5.1 Structure and indices of power plants. European part of Russia's UPS, 2010. (The table is compiled on the basis of [66] and the scientific reports of ESI SB RAS for 2006–2008)

No.	Indices	HPP	NPP	CPP on coal	CPP on gas STI	CPP on gas CCI	CGPP on coal	CGPP on gas
1.	Share in European part of UPS (%)	12.0	15.0	10.0	22.0	4.0	2.0	35.0
2.	Net efficiency (η)	–	0.33	0.35	0.39	0.52	?	?
3.	Service life (T_{ser}), years	100	50	30	30	30	30	30
4.	Specific investments (k), $/kW	2,200	1,650	1,200	950	800	1,500	1,200
5.	Annual fixed costs (μ), share	0.015	0.065	0.075	0.070	0.070	0.10	0.10
6.	Annual fixed costs (FC), $/kW/year	33.00	107.25	90.00	66.50	56.00	150.0	120.0
7.	Fuel price (P_f), $/tce	–	18.83	45.00	60.00	60.00	45.00	60.00
8.	Fuel component in costs (C_f), $/kWh	–	0.0070	0.0158	0.0189	0.0142	?	?

The economic indices are given in US$ 2005, which is convenient in many respects, in particular for their comparison with the similar indices abroad.

The fixed costs (FC) in line 6 are defined as a share (line 5) of the specific capital investments (line 4). The fuel component of costs (line 8) is calculated by the formula:

$$C_f = \frac{P_f}{8150\eta}, \tag{5.9}$$

where the efficiency η (line 2) is taken with electricity consumption of power plant auxiliaries. Hence, the fuel component is determined for the "net" electric power supplied from a power plant to the system (delivered to the market).

It should be underlined that the fuel component for NPP and CPP is taken as constant and does not depend on the volumes and conditions of electricity production. It is a frequent case in energy and economic calculations, especially in the analysis of EPS perspective development. It corresponds to the so-called linearization of cost structure (see, for example, [3]). Then expression (3.1) of the total costs for the short-run period (a year) can be written in the following form:

$$TC = FC + AVC \cdot Q, \tag{5.10}$$

where Q is the production volume for the considered period and $AVC = \text{const}$.

From expression (5.10), follow the interesting relations [3]:

$$ATC = FC/Q + AVC, \tag{5.11}$$

$$AFC = FC/Q \tag{5.12}$$

$$MC = \frac{dTC}{dQ} = AVC = const, \tag{5.13}$$

$$MC < ATC. \tag{5.14}$$

The relations mean that with the linearized structure of costs of a power plant (firm, company):

1. The curve of marginal costs (*MC*) and correspondingly the supply curve (*S*) will be a horizontal line (the curve *S* will convert to the vertical line, when the maximum possible production volume Q_m is reached).
2. If the power plant enters the market with such a supply curve and the equilibrium with the demand is on its horizontal section ($P = MC$), the power plant will not recover its fixed costs and go to smash.

The indicated consequences of such "linearization" should be taken into account in the further analysis of costs for different types of power plants. By analogy with [3] we consider now the impact of *utilization* of power plants during a year on their costs.

The data of Table 5.1 can be used to calculate average total costs (*ATC*) of HPPs, NPPs, and CPPs at different annual number of their capacity utilization hours (h). For simplification, the variable costs of NPPs and CPPs are assumed to consist of only fuel costs ($AVC = C_f$), and those of HPPs are in general equal to zero. Since the fixed costs *FC* (line 6) are expressed in $/kW/year and the variable (fuel) costs $C_f = AVC$ (in $/kWh), the total costs can be expressed by the formula:

$$ATC = FC/h + AVC, \tag{5.15}$$

which is similar to expression (5.11).

In accordance with formula (5.15) the average total costs for HPPs, NPPs, and CPPs were calculated based on the data of Table 5.1 as a function of the annual number of hours of their capacity utilization (h). They are presented in Table 5.2.

Table 5.2 Average total costs of power plants at different utilization of their capacity during a year, ¢/kWh

h, h/year	HPP	NPP	CPP on coal	CPP on gas STI	CPP on gas CCI
1,000	3.30	11.43	10.58	8.54	7.02
2,000	1.65	6.06	6.08	5.22	4.22
3,000	1.17	4.28	4.58	4.11	3.29
4,000	0.83	3.38	3.83	3.55	2.82
5,000	0.66	2.85	3.32	3.22	2.54
6,000	0.55	2.49	3.08	3.00	2.35
7,000	0.47	2.28	2.87	2.84	2.22

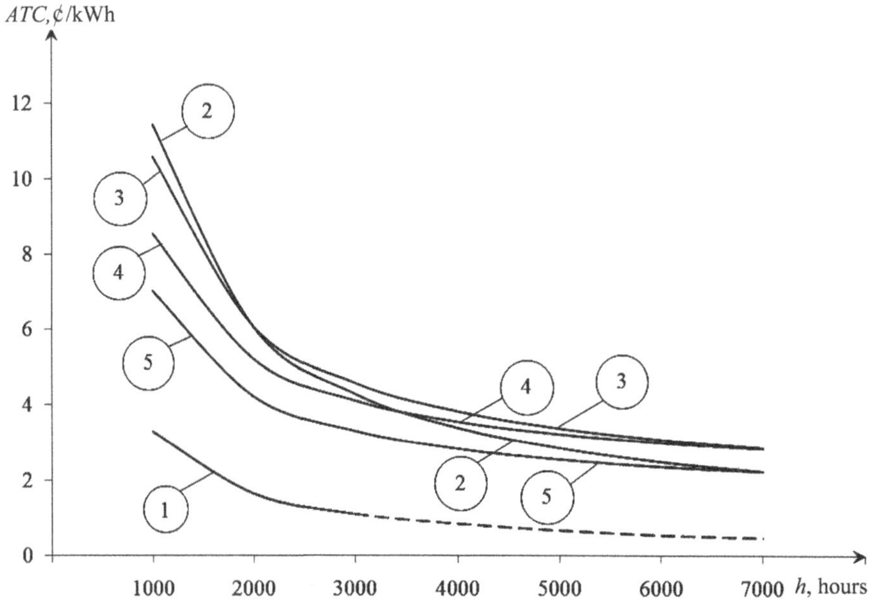

Fig. 5.1 Relations between the average total costs of power plants and the annual number of hours of their installed capacity utilization. *1* HPP; *2* NPP; *3* CPP on coal; *4* CPP on gas with STI; *5* CPP on gas with CCI

In Fig. 5.1, these dependences are presented graphically, allowing the following tendencies to be revealed:

1. The costs of HPPs are much lower than those of remaining power plants over the whole range of h. However, the annual hydro generation (and correspondingly the annual number of utilization hours) is known to be limited by the waterway conditions and in addition is random.
   Therefore, after $h = 3{,}000$ h their costs are shown by the dashed lines. They can be different if water inflow is low. HPPs, in principle, should be used in the peak zone of daily load curves of consumers within the limits of the volume of water inflow in the specific year.
2. The costs of nuclear generation are the largest at $h < 2{,}000$ h, but become lower than the costs of condensing generation on fossil fuel (except for CPP with CCI on gas) at $h > 4{,}000$ h. Hence, NPPs should be used to meet base loads, if possible.
3. The costs of condensing generation on gas with CCI are lower than the costs of remaining CPPs and NPPs over the whole range of h. Unfortunately, by the considered 2010 the share of CPPs with CCIs will remain relatively small.
4. The costs of condensing generation on coal *exceed* the costs of that on gas over the whole range of h; however, at $h > 3{,}000$–$4{,}000$ h they become close to the

costs of condensing generation with STI on gas. Therefore, it is expedient to use CPPs on gas to meet peak (in parallel with HPPs) and semi-peak loads and CPPs on coal to meet base (in parallel with NPPs) and semi-peak loads. In principle, the loading of CPPs on coal and gas can be approximately the same (especially if account is taken of the fact that the fuel costs for CPPs on coal are, on the contrary, lower than for CPPs on gas with STI) (Table 5.1).

The considered relations and the trends describe only the general impact of the *intra-year* operating conditions of power plants on their average costs. The actual relations may happen to be more complicated especially for CPPs and the more so for CGPPs that have not been treated yet. In fact, it is necessary to analyze the curves of *short-run* average costs (total, fixed, and variable) for different types of power plants and also the curves of marginal costs (*MC*) and the supply curves (*S*), with which the power plants will enter the short-run (competitive) market of electricity. Now we proceed to this analysis.

The short-run and, to be more concrete, *annual costs* of power plants will be analyzed as a function of their annual electricity generation under the following assumptions:

- The installed capacity of each power plant is taken to be equal to 1 GW (1,000 MW). This will make it possible to switch further to the total capacity of power plants of the considered type in the European part of UPS.
- The load among the power plant units is distributed optimally (taking account of the expenses on their startup and shutdown). In this case the efficiency of the whole power plant (and its *AVC*) will be leveled within the range of possible annual electricity generation.
- The intra-year load distribution of EPS among different types of power plants is optimized by using the commonly accepted technique, i.e., with optimization of "instantaneous" (hourly and daily) operating conditions of EPS, optimal distribution of the annual flow of HPP by season of the year and day of the week, etc. This will necessitate consideration (to a certain extent) of the mentioned trends in using different types of power plants in the base, peak, and semi-peak zones.
- The characteristics of power plant costs do not depend on the specific type (model) of electricity market organization. The market models will influence the price formation, which will be addressed in the next section.
- The annual costs of power plants (both variable and total) are to a great extent uncertain (depending on the intra-year operation of power plants, their structure in EPS, etc.). When this uncertainty is particularly high, the cost curves will be set by some zone or by several variants.
- During the year a certain part of installed capacities of power plants is under repair or in reserve, which decreases their production in comparison with the theoretically maximum possible one. In this context the maximum annual number of capacity utilization hours of power plants (in particular, NPPs) during their base load operation will be taken as $h_{max} = 7{,}000$ h.

5.3.2 Short-Run Costs of HPPs

Relations between the annual costs of HPPs and their annual production are simpler than of the other types of power plants, since the variable costs of HPPs (*AVC*) are practically equal to zero. The average total costs (*ATC*) will consist of only the fixed costs ($ATC = AFC$). The marginal costs of HPPs will correspondingly be equal to zero ($MC=0$) over the whole range of possible annual hydro production.

At the same time, the annual hydro production is determined by water inflow at different years and is, therefore, a random variable. This fact brings about one more peculiarity in the characteristics of the annual costs of HPPs.

As was already noted, it is expedient to use hydro generation for meeting peak daily loads for consumers. The intra-year operation of HPPs will be in this case most nonuniform; however, it does not essentially deteriorate the energy indices of HPPs because of two reasons:

1. The efficiency characteristics of HPP units are flat enough over a wide range of their capacities (water flow rates).
2. The number of units at HPPs is usually rather large (5–10 and sometimes even tens of units) and there are virtually no expenses on the startup and shutdown of units.

Therefore, the optimal choice of the number of operating units will guarantee high efficiency (close to 95%) almost over the whole range of instantaneous (hourly) capacities of HPPs.

Figure 5.2 presents dependences of the short-run costs of the 1 GW HPP. Based on the fact that practically HPPs have no variable costs, the form of presented dependences completely corresponds to the "linearized structure of costs," i.e., to the above considered expressions (5.10)–(5.14), in which $AVC=0$.

The average total costs of HPPs consisting of the fixed costs only ($ATC = AFC$) decrease with the increasing annual production Q. For HPPs with the installed

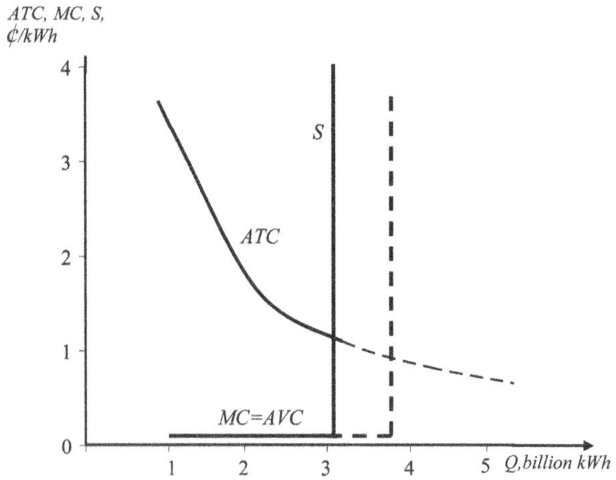

Fig. 5.2 Short-run (annual) costs of HPP with a capacity of 1 GW

capacity 1 GW, the curve for *ATC* is fully identical to curve 1 in Fig. 5.1, with the decrease in abscissa scale by 3 orders of magnitude ($Q=1$ GW × *h*, billion kWh).

According to expression (5.13) the marginal costs of HPPs are equal to zero ($MC=AVC=0$) at any Q (in Fig. 5.2 they are shown to be close to zero). In the competitive wholesale electricity market (Models 3 and 4) the supply curve of HPPs (*S*) on the section before the maximum possible annual production Q_m will also be equal to zero ($S=MC=0$), and at Q_m will turn into a vertical range. So far as Q_m is a random variable, in Fig. 5.2 the curve *S* is conventionally shown by the solid line at $Q_m=3$ billion kWh and by the dashed line (that will shift) at $Q_m>3$ billion kWh.

Comparison of Fig. 5.2 with Fig. 3.2 in Chap. 3 which was plotted for "typical" firms treated in the microeconomics proves that they are basically different:

- The curve *ATC* for HPPs continuously decreases, reaching its minimum at the maximum annual production Q_m.
- The supply curve *S* consists of only the horizontal (zero) and the vertical sections.
- Over the whole range of Q the marginal costs are lower than the average total costs ($MC < ATC$), i.e., inequality (5.14) is satisfied.

Hence, HPPs definitely cannot be referred to the category of "typical" firms participating in the "classical" competitive markets.

At the same time the relations of HPP costs shown in Fig. 5.2 are quite objective (real). Section 5.4 will illustrate how they can be applied to determine aggregate electricity generation costs for EPS as a whole (in particular, for the vertically integrated monopoly company).

5.3.3 Short-Run Costs of NPPs

Unlike HPPs it is advisable to use NPPs to meet base loads. Their maximum annual generation will depend on the downtime of NPP blocks in scheduled and emergency repairs (including refueling). As was already noted, the maximum possible number of annual capacity utilization hours will be taken for certainty equal to 7,000 h. In principle, it can be lower, if at some periods of time NPPs do not operate at maximum capacity (are unloaded).

Consumption of nuclear fuel by NPPs is usually assumed to be directly proportional to electricity production (the author has not met nonlinear characteristics of NPPs). Therefore, "linearization of cost structure" can be fairly taken for NPPs. This assumption was applied earlier to compile Tables 5.1 and 5.2, plot Fig. 5.1, and write expressions (5.10)–(5.12). As a result, the analysis of the short-run costs of NPPs becomes much simpler.

Figure 5.3 presents the average total cost (*ATC*), average variable cost (*AVC*), and marginal cost (*MC*) as well as the supply curve (*S*) for NPPs with a capacity of 1 GW. They were constructed on the basis of the data of Tables 5.1 and 5.2. The

Fig. 5.3 Short-run costs of NPPs with a capacity of 1 GW

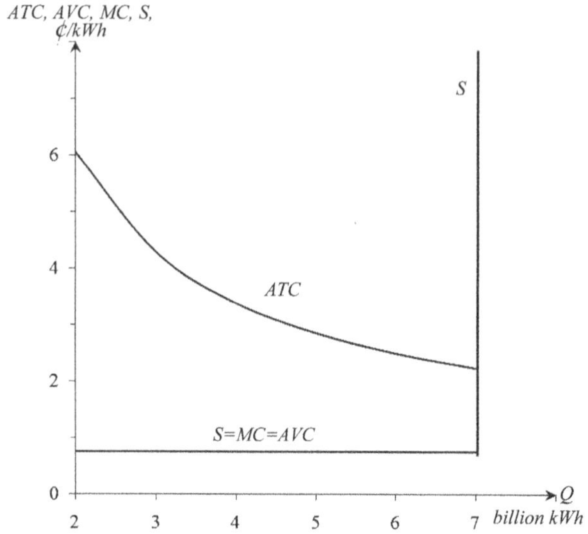

curve for *ATC* is identical to curve 2 in Fig. 3.6, considering the scale and the ranges of annual production *Q*.

The general shape of curves in Fig. 5.3 is similar to that of curves for HPPs in Fig. 5.2. The only difference is that the variable and marginal costs are not equal to zero, but 0.7 ¢/kWh. The average total costs (*ATC*) steadily decrease, reaching the minimum at the maximum annual production Q_m=7 billion kWh. The supply curve (*S*), with which NPPs may appear in the *competitive* market, again has only a horizontal and a vertical section.

Hence, NPPs and HPPs radically differ in their economic properties from the "typical" firms addressed in microeconomics. As will be shown later, this fact substantially influences their participation in the competitive markets with free prices (Models 3 and 4).

5.3.4 Short-Run Costs of CPPs on Fossil Fuel

The short-run (annual) costs of CPPs require a more complicated analysis than HPPs and NPPs, since the assumption on the constancy of their variable costs (*AVC*) and correspondingly on the "linearization of cost structure" is not obvious now. Nonlinearity of the hourly (instantaneous) cost characteristics of CPP units is well known and should undoubtedly be taken into consideration while optimizing the EPS operation conditions within a day. Besides, additional fuel expenses for startup and shutdown of CPP units are rather large and should also be considered (as to HPPs such expenses are virtually insignificant and the NPP blocks are shut down only for repair).

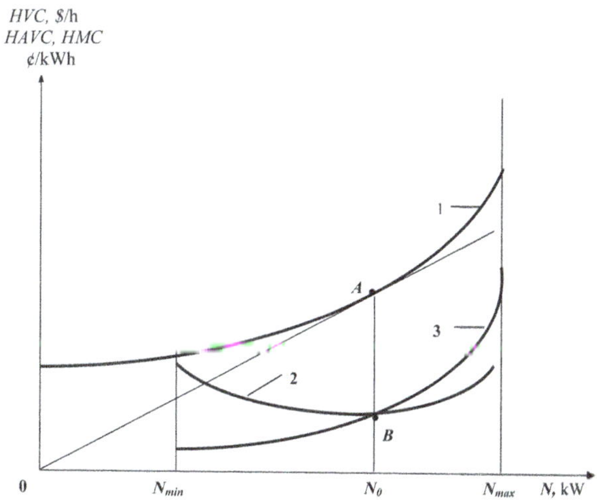

Fig. 5.4 Relationships between the hourly costs of CPP unit and the capacity. *1* full hourly production costs *HVC*, *2 HAVC*, *3 HMC*

Though in this section we are finally interested in the relations between the *average short-run costs* and *the annual CCP generation*, it makes sense to start analysis with the *hourly or instantaneous characteristics*. In order to distinguish them from the short-run (annual) costs we will put before them the letter H (hourly), as is done in Sect. 5.1.

Figure 5.4 presents a fuel cost characteristic of a CPP unit (curve 1).[1] The fuel consumption is assumed to be multiplied by its price and hence it can be interpreted as the relationship between the hourly variable costs (*HVC*) and the unit loading (working capacity *N*). Practically always after commissioning, the units have a minimum admissible load ($N_{\min}$), below which they should be stopped.

In curve 1 of interest is the point *A*, at which it touches the straight line drawn from the coordinate origin. At this point the hourly average variable costs (*HAVC*) are equal to the marginal costs (*HMC*):

$$HAVC = \frac{HVC}{N} = HMC = \frac{dHVC}{dN}. \tag{5.16}$$

Curve 2 is the relation between the average (hourly) variable costs and the capacity. It reaches the minimum at point *B*, which corresponds to the capacity N_0. Curve 3 of the hourly marginal costs intersects it at this point. If the capacity is lower than N_0, the curve *HMC* is below the curve *HAVC*, i.e., while the marginal costs are lower than the average ones, the latter decrease. On the section to the right of N_0 the marginal costs exceed the average ones and the latter rise up to the maximum capacity $N_{\max}$.

Note that the hourly marginal costs are completely analogous to the characteristics of incremental rates that were extensively applied to optimize daily operating

[1] Characteristics of gas turbines have a different shape (convex upwards), however they will not be analyzed, since in Russia they are applied virtually only in the combined cycle.

conditions of EPS in Russia (see, for example, [67, 68]). Power plants should enter the spot markets, if organized, with the same costs, as is supposed in the conceptions of the competitive markets.

The curves of average (*HAVC*) and marginal (*HMC*) costs in Fig. 5.4 are similar to the curves of costs for the "typical" firms presented in Fig. 3.2 in Sect. 3.1. Point *B* at the intersection of the curves *HAVC* and *HMC* in Fig. 5.4 corresponds to point *A* in Fig. 3.2.

At the same time the curves of costs in Fig. 3.2 and Fig. 5.4 drastically differ:

- Figure 3.2 presents the relationships between the *short-run* (annual, to be more concrete) costs and the *annual* production *Q*, and Fig. 5.4 presents the relationships between the *hourly* costs and the working *capacity* of the unit. Complete analogy should be achieved by transition from the instantaneous (hourly) capacities to the annual electricity generation, which proves to be complicated enough and uncertain.
- Figure 5.4 is constructed for only *one unit* of CPP and Fig. 3.2 for the *firm* (in our case, a power plant) as a whole. Therefore, in Fig. 5.4 there are naturally no fixed costs (*AFC*) of the whole power plant and its total costs (*ATC*).

In principle, the *hourly* (instantaneous) characteristics of costs for the *whole CPP* can be derived based on the characteristics of units, though it will cause difficulties in consideration of additional fuel consumption for unit startup and shutdown. In particular, the characteristics of incremental rates of fuel consumption and cost (i.e., actually hourly marginal cost) were derived for CPP by some or other method and applied in the 1950s–1980s of the last century in Russia for optimal distribution of EPS load among power plants [67, 68]. In parallel, the aggregate characteristics for the whole EPS were obtained and they were similar to the aggregate supply curve of the industry shown in Fig. 3.3 in Sect. 3.1.

The shape of curves of hourly costs for CPP consisting of several identical units will be the same as the shape of curves in Fig. 5.4, assuming that the *units are in operation*. Under such an assumption one must only change the scales of the values on the abscissas and ordinates, multiplying them by the number of operating units (the load among the operating units should be distributed uniformly).

Meanwhile, when optimizing operating conditions of EPS for a day or a whole week, shutdown of one or several CPP units at nights or for weekends is, as a rule, economically feasible. Expediency of shutdowns depends on many factors: fuel consumption for startups and shutdowns, daily and weekly load curves of consumers, general mix (structure) of different power plants in EPS, etc. It proves to be impossible to derive characteristics of *hourly* costs of CPP (as a function of capacity) at shutdown of units. One can only derive the relationships between:

- The *daily* costs and the *daily* CPP generation (or average daily capacity), if the units are shut down for a night, or
- The *weekly* costs and the *weekly* electricity generation by CPP (or its average weekly capacity), if the units are shut down for weekends.

The relationships for the daily or weekly costs of CPP will not be quite certain any more, as in the case for one unit in Fig. 5.4 or for several units in operation. Their form will depend on the factors mentioned above, mainly on the configuration of consumer load curves and the composition of the other power plants in EPS. They can be constructed by a rather wide series of optimization calculations of the daily or weekly operating conditions for a specific EPS. For one and the same CPP such characteristics will differ depending on the mentioned factors as well as on a season of the year. Still some *common features* can be indicated:

- The daily and weekly costs, by analogy with the hourly ones, will represent only *variable* costs of CPP and not include fixed costs of the power plant.
- The curves of *average* daily and weekly costs of CPP will have a U-shaped form (as the curve for *HAVC* in Fig. 5.4). However, the region of minimum cost values will be "extended" owing to optimization of unit commitment and optimal distribution of total CPP load among its units.
- In the regions of *minimum* and *maximum* daily (or weekly) CPP generation, the values of average costs *will increase* because of the general nonlinearity of the cost characteristic of the unit (curve 1 in Fig. 5.4).

These features will be seen in the characteristics of the annual (properly speaking, short-run) *costs* of CPP that are of concern for us and will be considered below.

The relationships between the average (*ATC*, *AVC*) and the marginal (*MC*) costs of CPPs and their annual production will be more sophisticated than the considered relationships for HPPs (Fig. 5.2), whose variable costs are virtually equal to zero, and for NPPs (Fig. 5.3), whose variable costs may be treated as fixed. Nonlinearity of the cost characteristics of CPP units (Fig. 5.4) will be displayed in the annual costs, though to a considerably lower extent. It would be apparently methodologically incorrect to neglect this nonlinearity.

The average fixed costs (*AFC*) will naturally depend only on the annual CPP generation Q. The variable (*AVC*) and total (*ATC*) costs will be determined by the intra-year operating conditions of CPP. The effect of daily and weekly operating conditions has already been considered above and expressed primarily in "leveling" (decrease in nonlinearity) of the curves for variable costs because of shutdown of some CPP units and optimal distribution of the total load of CPP among them. Besides, for the intra-year operating conditions of CPP one should take into account:

1. *Load changes* of EPS consumers *by season of the year*. In the seasons of maximum load, CPPs will be loaded to a greater extent and in some hours up to the maximum available capacity.
2. *Location of operating capacity reserves* of EPS at CPPs. It will surely depend on the general structure of generation capacities of EPS and can be different in different EPSs and in different seasons of the year.
3. *Outages* of part of power plant units in the scheduled and emergency repairs both at CPPs and other types of power plants.
4. *Optimization* of EPS operating conditions in all time intervals (cycles) of the year and with any unit commitment. This will provide the most favorable conditions for CPP operation (with the least specific fuel consumption).

The indicated factors allow the following statement. First, for the larger part of the year (hours of the year) CPP will operate with intermediate (not maximum and not minimum) loads that correspond to the optimal conditions of EPS as a whole, including the optimal unit commitment and location of capacity reserves. It leads to even greater "leveling" of the curves of *annual* variable costs of CPPs as against the considered daily and weekly costs. At the same time during the hours of the year, when CPPs operate with maximum or, on the contrary, minimum loads, the specific fuel consumption and the costs of CPP will grow.

Second, the characteristics of annual costs of CPPs will be ambiguous (uncertain) to an even greater extent than the characteristics of daily and weekly costs. Their specific form can be shown only approximately with indication of solely general revealed trends.

From the above, Fig. 5.5 presents a possible form of the short-run costs of a CPP *on gas with* STIs of the total capacity of 1 GW. The short-run costs are based on the characteristics of such CPP with the "linearized structure of costs" that are shown in Tables 5.1 and 5.2. In order to reduce the sizes in Fig. 5.5, the range of annual CPP generation Q is assumed to be 2–7 billion kWh, which corresponds to $h=2{,}000$–$7{,}000$ h/year.

With some degree of conventionality (in an expert way) it is assumed that in the range $Q=3$–5 billion kWh CPP has the most favorable conditions, and the variable costs (*AVC*) will equal 1.89 ¢/kWh. At $Q<3$ billion kWh and $Q>5$ billion kWh, they increase.

Correspondingly, the marginal costs (*MC*) will:

- Be equal to the variable costs ($MC=AVC$) in the range $Q=3$–5 billion kWh;
- Exceed the latter ($MC>AVC$) at $Q>5$ billion kWh; and
- On the contrary, be lower ($MC<ATC$) at $Q<3$ billion kWh.

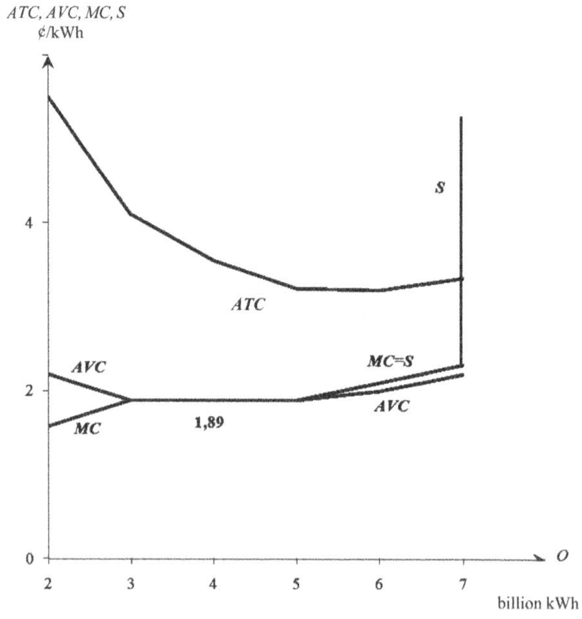

Fig. 5.5 Short-run costs of CPP on gas with STI of 1 GW

The supply curve of CPP (*S*) will coincide with the curve *MC* at $Q>5$ billion kWh and will turn to a vertical range at the maximum possible production $Q_m=7$ billion kWh.

The shown shape of the curve of variable costs *AVC* will influence the curve of total costs *ATC* that will slow down its fall at $Q>5$ billion kWh and even somewhat increase at the maximum production $Q_m=7$ billion kWh.

Thus, the shape of the curves for the short-run (annual) costs of CPPs may be expected to differ to some extent from the curves of costs for HPPs and NPPs, approaching the form of costs of the "typical" firms (Fig. 3.2). However, it should be noted that:

- The U-shaped form of the curve *AVC* is manifested very poorly, and
- The curve *MC* cannot intersect the curve *ATC*, i.e., the supply curve *S* will not guarantee payback of the fixed costs *AFC*.

This fact will influence the CPP participation in the *competitive* electricity market.

Despite the conventionality of Fig. 5.5 associated with the mentioned earlier uncertainty and difficulties in construction of the curves for the short-run costs of CPP, a similar shape of the curves may be expected for the other types of CPPs (on coal and gas with CCI).

5.3.5 Short-Run Costs of CGPPs

An analysis of the short-run (annual) costs of CGPPs proves to be the most complex and ambiguous. Our concern is with the CGPP costs for electricity production that determine their participation (together with other power plants) in the electricity market. However, in parallel with electricity, CGPPs produce heat (it means thermal energy, for short). This fact drastically complicates the analysis.

Combined heat and electricity production is known to result in savings of investments and costs as against their *separate* production by boiler plants and CPPs. The economic efficiency of concrete CGPPs is determined just by comparing the costs of separate and combined energy supply. It depends on many factors and conditions: concentration of heat loads, fuel kind and cost, technologies used at CGPPs, boiler plants and CPPs, etc. A great variety of these conditions over the territory of the country and settlements (including industrial enterprises in them) also stipulate highly diverse thermal and electric capacities of CGPPs and their other parameters and technical and economic indices. In fact, each CGPP is individual in its parameters, equipment, and indices.

Nevertheless, it can be assumed that CGPP (if built) provides a certain *saving* of expenses for *combined production* of electricity and heat. We will proceed from this assumption, though in practice it cannot be always satisfied, if the actual conditions of CGPP operation essentially differ from those envisaged in its designing.

The main difficulty in the determination of CGPP costs on electricity production is *to distribute the total costs* for combined electricity and heat production *between these two kinds of energy*. Factually, there is *no quite an objective method* or way for distribution of these expenses. The expediency to attribute the obtained saving (i.e., price reduction) to thermal or electric energy is determined by specific conditions subjectively to a great extent. In the USSR, for example, it was common practice to attribute the whole saving to electric energy (to consider that the combined production at CGPPs decreases specific fuel consumption for electricity generation). At that time in Federal Republic of Germany the saving, on the contrary, was attributed to heat to attract (to extend the circle of) consumers of thermal energy. At present this method is being applied more and more in Russia. In principle this saving can be distributed between both kinds of energy in some proportion (in particular, proportionally to the volumes of electricity and heat production).

Thus, the CGPP costs for electricity production will depend on the method of distributing the savings achieved owing to combined production between these kinds of energy. Moreover, both the saving of variable (fuel) costs and the fixed costs of CGPPs should be distributed, in particular, in its total capital investments. This fact naturally brings about uncertainty when the considered costs are determined.

Besides, CGPPs are characterized by sets of other specific features stipulating diversity of their parameters and economic indices:

- An individual character of each CGPP in the volumes and types of heat loads (steam of different temperatures, hot water for heating, and hot water supply). Therefore, diverse combinations of different types of extraction turbines (with backpressure, with one or several steam extractions), gas turbine units, and even purely condensing turbines can be installed at CGPPs. Hence, combinations of boiler facilities and the total CGPP capacities for thermal and electric energy production will also be different.
- Possible condensing generation (without heat production) and thermal energy delivery through pressure-reducing and desuperheating stations (PRDS) besides the turbines (without electricity production). For condensing generation at CGPPs the specific fuel consumption (and fuel costs) per 1 kWh of electricity proves to be much higher than for combined production and even somewhat higher than for CPPs. Heat supply through PRDS also increases specific fuel consumption. However, this type of generation will not be dealt with, since we are interested only in the costs for electricity production.
- The fundamental distinction of operating conditions and economic indices of the majority of CGPPs in the heating (winter period in Russia) and nonheating seasons. It is dictated by a large (as a rule) share of CGPP loads for heating of cities and settlements. In the nonheating seasons CGPPs cover only industrial thermal loads and hot water supply. Electricity, therewith, is generated basically in the condensing (less economical) conditions. The length of the heating season depends on climatic conditions and varies over the territory of the country (and to some extent, from year to year).
- Difference in fuel kinds and cost, conditions of water supply to CGPPs, and some other factors.

Because of the indicated peculiarities (and uncertainties) the characteristics of the short-run costs of CGPPs can be derived only on the basis of rather numerous assumptions that characterize specific conditions of their construction and operation. We will try to do this *on the example of some "typical" gas-fired CGPPs* with steam turbines that prevail in the European part of Russia. The indices of such a CGPP that are presented in Table 5.1 are taken as the base:

- Specific capital investments—\$1,200/kW$_e$
- Annual fixed costs (*FC*)—\$120/kW$_e$/year (or 10% of the specific capital investments)
- Fuel price—\$60/tce

Impose the following concrete assumptions:

1. The installed electric capacity of the CGPP (the rated capacity of turbine generators) is equal to 1 GW (the same as for the other considered types of power plants). The maximum capacity of the CGPP in thermal energy supply is also equal to 1 GW (860 Gcal/h). As long as these capacities are equal, the specific capital investments and the annual fixed costs given above will be distributed between the thermal and electric energy production in equal parts. Then the fixed costs of our CGPP *for electricity production* will amount to \$60/kW$_e$/year.
2. The length of the heating season with combined production of electricity and heat is taken to be equal to 4,000–4,500 h/year.
   Let us consider three methods of distributing the variable (fuel) costs of a CGPP between heat and electricity:

   a. *All the saving* owing to combined production applies to electric energy that will be the cheapest in this case. In the expert way it will be assumed here that

      - The specific consumption of coal equivalent for electricity generation will make up $q_1 = 200$ gce/kWh, and
      - The average variable costs (at the fuel cost of \$60/tce) $AVC_1 =$ 1.20 ¢/kWh.

   b. *All the saving* applies to thermal energy. The electric energy will naturally be more expensive. Setting that

      - The specific fuel consumption will be in this case approximately the same as for CPPs $q_2 = 315$ gce/kWh, then
      - The average variable costs will be equal to $AVC_2 = 1.89$ ¢/kWh.

   c. An *intermediate* method, for which we assume that

      - The specific fuel consumption $q_3 = 250$ gce/kWh and
      - The average variable costs $AVC_3 = 1.50$ ¢/kWh.

   These three methods and the values of variable costs obtained for them reflect the mentioned uncertainty in distribution of the saving owing to combined production between electric and thermal energy. The assumed figures of specific fuel consumption should be considered as conditional. For specific CGPPs they can essentially differ.

3. The maximum electricity production by the considered CGPP is assumed to be 7 billion kWh, which corresponds to utilization of its installed electric capacity of 7,000 h/year. An additional electricity generation (above its combined production during the heating period) will be provided mainly by the condensing generation (though combined production of heat for industrial needs and hot water supply is possible), with larger specific fuel consumption and variable costs. Assume that their following values for the condensing generation are:

   a. Specific consumption of coal equivalent $q=330$ gce/kWh;
   b. Average variable costs $AVC=1.98$ ¢/kWh.

Based on the above assumptions Table 5.3 presents the average costs calculated for the studied CGPP at different annual electricity production.

The fixed costs (*AFC*) were calculated by formula (5.12):

$$AFC = FC/Q, \tag{5.12}$$

where $FC = \$60 \times 10^6/\text{year}$.

The variable costs (*AVC*) at combined production are given for all the three methods of distributing the achieved saving. Similar to the condensing generation they are taken constant, independent of the electricity production volumes. On the one hand, it is due to the general uncertainty of cost characteristics that is provoked by the diversity of the intra-year (daily, weekly, seasonal) operating conditions of CGPPs with different proportions in generation of electric and thermal energy and different unit commitments. On the other hand, the author has not managed to find materials that could help quantify relationships between the CGPP specific costs and the annual electricity production.

The curves of the CGPP costs for electricity production are plotted in Fig. 5.6 in accordance with the data of Table 5.3. The curves for the variable (*AVC*) and total (*ATC*) costs have "breaking points" at transition from the combined to condensing generation. As far as the variable costs on these two sections are constant, they are also the marginal costs ($MC = AVC$), and the latter are lower throughout than the total costs ($MC<ATC$).

Table 5.3 Average costs of electricity production by 1 GW_e CGPP, ¢/kWh

Annual electricity production *Q*, billion kWh	Fixed costs *AFC*	Variable costs *AVC*				Total costs *ATC*[a]
		Combined generation			Condensing generation	
		1	2	3		
1.0	6.00	1.20	1.89	1.50		7.50
2.0	3.00	1.20	1.89	1.50		4.50
3.0	2.00	1.20	1.89	1.50		3.50
4.0	1.50	1.20	1.89	1.50		3.00
5.0	1.20				1.98	3.18
6.0	1.00				1.98	2.98
7.0	0.86				1.98	2.84

[a] For the third (intermediate) method of saving distribution from combined production.

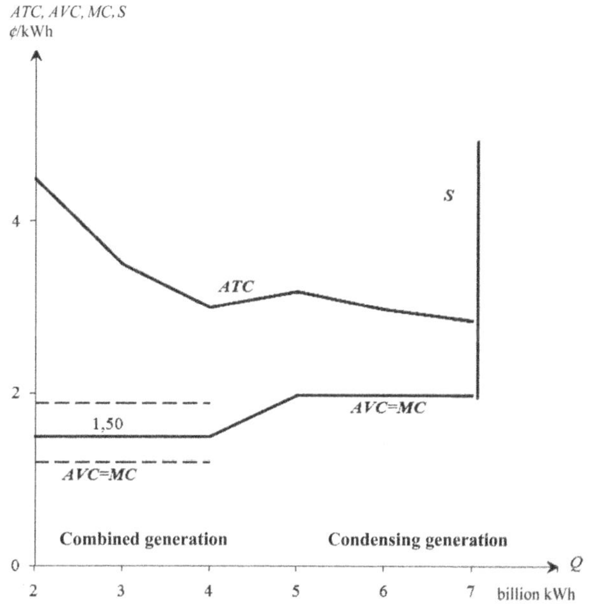

Fig. 5.6 Short-run costs for electricity production by gas-fired CGPP of 1 GW with STI

On the whole, the curves for the short-run costs of CGPPs (like for the other types of power plants) substantially differ from the curves for costs of "typical" firms (Fig. 3.2).

The uncertainty of the CGPP costs for combined electricity production should be noted once again (the dashed lines in Fig. 5.6). Validity (and even possibility) of CGPP participation in the *competitive electricity* market will depend on prices (or tariffs) for *thermal* energy. A particularly complex situation will take place when the competitive (with free prices) heat market is organized along with the competitive electricity market. In this case CGPP will participate in both markets and must coordinate its price bids (or contracts) for thermal and electric energy. If there is no heat market and the tariffs for thermal energy supplied from CGPP are regulated, the saving owing to combined production will be attributed to electric energy on "leftovers." In this case the profitability (or nonprofitability) of CGPP operation will be determined by the prices formed in the electricity market.

Thus, the CGPP participation in the competitive electricity market turns out to be more complicated than of the other power plants. It is much simpler when the prices of both heat and electricity are regulated in a mutually coordinated way for CGPP as a whole.

5.3.6 Comparison of Short-Run Costs of Power Plants with Costs of "Typical" Firms

Remind that we deal with the costs of firms (and power plants) at the *short-run* period, during which the firm's production capacity does not change (is fixed). For certainty the length of the period is taken to be equal to 1 year.

The analysis of Figs 5.2, 5.3, 5.5, and 5.6 and their comparison with Fig. 3.2 in Sect. 3.1 makes it possible to reveal the following basic distinctions between the curves of power plant costs and the similar curves of the "typical" firms treated in microeconomics:

1. *The curves of variable (AVC) and total (ATC) costs of power plants do not have a U-shaped form* (i.e., the minimum), as is assumed for the "typical" firms. The average total costs (*ATC*) reach the minimum at the maximum possible annual generation of power plants.
   The CPP costs may be an exception, because they somewhat increase at the annual number of their installed capacity utilization hours above 6,000. However, in the absence of power shortages in EPS, location of part of necessary reserves at CPP, and optimization of power plant operating conditions each hour, the real CPP loading will not reach such values. Hence, the total costs (*ATC*) for CPP may be treated as continuously decreasing with the increasing annual production.
   This specific feature can be explained by one of the EPS properties addressed in Chap. 2, i.e., high mechanization, automation, and even robotization (at NPPs) of electricity production, transmission, and distribution processes. All the processing lines and nodes of power plants are designed for their total *installed* capacity. As a result, the power plants have:

   a. A high share of fixed costs, whose specific value decreases with the increasing annual electricity generation.
   b. Only the managerial staff, duty, and maintenance personnel, whose number does not virtually depend on the actual electricity production. These personnel wages also represent the fixed costs.
   c. The variable costs consisting practically of fuel costs. And thanks to optimization of unit commitment of power plants and EPS operation conditions in general, their specific values are slightly dependent on the annual output of CPPs and CGPPs. NPPs usually operate under uniform base conditions and HPPs have no variable costs at all.

2. *The marginal costs (MC) are as a rule equal to variable costs (AVC) and always remain lower than the total costs (ATC)*. The *marginal costs* of the "typical" firms (Fig. 3.2) exceed the variable and total costs on the rising branches of the U-shaped curves of *AVC* and *ATC*.
   It means that the power plants cannot participate in the market with their marginal costs (as is assumed for the "typical" firms), because their fixed costs will not be compensated for. *Power plants must participate in the market* (submit price bids or make contracts) *based on their total costs (ATC)*.
3. *Characteristics of short-run costs of power plants are to a great extent uncertain.* This is explained by many factors and specific features:

   a. The annual electricity production at HPPs is uncertain (random) because it depends on water inflow.
   b. Uncertainty of the annual costs of CPPs is caused by the diversity of their within-year (daily, weekly, seasonal) operating conditions that depend on the general structure (mix) of generation capacities of EPSs.

c. As to CGPPs along with the variety of the within-year conditions the uncertainty is increased by outdoor temperature fluctuations and the diversity (individuality) of specific CGPPs in their parameters and equipment mix.

The characteristics of short-run costs of NPPs only prove to be certain enough owing to their uniform (base) conditions of operation.

The indicated distinctions (features) of cost characteristics of power plants present, as was shown in the previous section, difficulties (and even barriers) in organizing the spot electricity market. They also influence the formation of the wholesale prices in the short-run market, which will be dealt with in the next section.

Box 12 Specific Features of Short-Run Costs of Power Plants

1. The short-run (annual) costs of power plants radically differ from the instantaneous (hourly) costs that are applied to optimize EPS operation conditions. The short-run costs include fixed costs and depend on the annual electricity production.
2. The curves of short-run costs of power plants *have no* U-shaped (with the minimum) form, as is assumed in the microeconomics for "typical" firms. The average total costs (*ATC*) reach the minimum at the *maximum possible* annual electricity production of the power plant. This is explained by a high level of mechanization and automation of electricity production processes, design of all power plant facilities for its full installed capacity, and practical independence of personnel number from the actual electricity production.
3. The variable costs (*AVC*) of CPPs (and NPPs as well) *consist* of fuel costs *only* (for HPPs they are equal to zero). As a rule, the marginal costs (*MC*) are equal to variable costs (*AVC*) and are *always lower* than the total costs (*ATC*). It means that if the price offered by power plants in the market corresponds to marginal costs (*MC*), as is supposed for the "typical" firms, then it will not compensate for the fixed costs (*AFC*).
   Thus, the power plants must participate in the market (conclude contracts) based on their *short-run* (annual) *total costs* (*ATC*).

5.4 Short-Run Costs of Generation Companies and Price Formation in the Wholesale Electricity Market

This section will make an analysis of *shot-run* (annual) costs of VICs and individual PGCs on the basis of cost curves of different types of power plants that were obtained in Sect. 5.3. The installed capacities of power plants (and overall companies) will be assumed to be fixed (invariable) and the annual outputs to be varying within some (realistic) limits. The cost characteristics of VICs and PGCs will be used to illustrate the price formation in the wholesale market for regulated (Models 1 and 2) and unregulated (Models 3 and 4) prices. In particular, the producer's surplus formation under unregulated prices will be presented. As in Sect. 5.3, cost characteristics will be shown by the example of the European section of Russia's UPS for 2010.

For VICs (Model 1) the costs will be considered only in the *sphere of generation*. These will then be supplemented by the costs in the other spheres to determine tariffs for consumers. However, these spheres will not be considered here. VIC includes naturally *all* power plants situated on the territory served by it. Additionally, it can exchange power (selling or buying) with neighboring VICs which will affect consumer tariffs; however here we will discuss only the costs of VIC generation.

PGCs may consist of one or several power plants. The case of one power plant has already been considered in Sect. 5.3. Therefore, here the analysis of the second, more general case will be made. Power plants that belong to one PGC can be of the same type or of various types and they can be spread over the territory of the country. For compatibility the assumption will be made that power plants of individual PGCs are situated on the territory of the considered VIC.

Cost characteristics of PGC do not depend on the model of electricity market organization. However, their participation in the wholesale market will be different under regulated (Model 2) and unregulated (Models 3 and 4) prices. Price (tariff) formation under its regulation will be practically the same as under monopoly VIC. For competitive markets it will be radically different.

5.4.1 *Short-Run Costs of VICs*

In the same manner as in Sect. 5.3, we will determine the relationships between average fixed (AFC_{VIC}), variable (AVC_{VIC}), and total (ATC_{VIC}) costs, expressed in ¢/kWh, and the annual output of the company Q_{VIC}, measured in billion kWh. Fixed installed capacity of VIC represents, naturally, a sum of installed capacities of power plants that belong to the VIC (for simplification power plants will be differentiated only by their type $i=1,\ldots,I$):

$$N_{\mathrm{VIC}} = \sum_{i=1}^{I} N_i. \qquad (5.17)$$

The annual electricity output of VIC will also be equal to a sum of annual outputs of individual types of power plants (Q_i), which can be characterized by the annual number of capacity utilization hours (h_i):

$$Q_{\mathrm{VIC}} = \sum_{i=1}^{I} Q_i = \sum_{i=1}^{I} h_i N_i. \qquad (5.18)$$

In order to construct the curves of VIC costs we will set several values of Q_{VIC} which are possible at its fixed installed capacity N_{VIC}. These values should be distributed among the outputs of individual types of power plants Q_i which the costs of these types will depend on. Such a distribution will be made on the assumption that the within-year operation of EPS (daily, weekly, and annual) is *optimized according to the criterion of minimum variable costs* of the entire VIC. As is known at optimization of EPS operation the most efficient types of power plants are loaded or utilized fully, while the remaining ones are loaded partially to supplement the former to achieve the required power and energy balance in the system. Taking into account the experience of Russian UPS optimization and calculations in Sect. 5.3 (Fig. 5.1) we can make the following assumptions.

- *Annual output* of HPPs, depending on hydrological conditions (with account taken of its optimal distribution among reservoirs by hour of the day and season of the year), is *used completely*. The number of HPP capacity utilization hours in this case is 3,000–4,000 h/year.
- *Available capacity* of NPPs which operate for basic load is also used completely. Their installed capacity is used nearly 7,000 h/year (as was assumed in Sect. 5.3).
- *Electricity cogenerated* by CGPPs (along with heat) is *also used completely*. Utilization of electric capacity at combined generation makes up 4,000–4,500 h/year.
- *The marginal sources in power balance of* EPS (VIC) will be CPPs and condensing generation of CGPPs. Their annual output will be determined by the shapes of daily and annual load curves of consumers and by the total annual electricity consumption.

These assumptions will be taken into account to determine annual output of individual types of power plants Q_i at a set electricity generation by the entire VIC Q_{VIC}.

Let us make an analysis and construct the curve of *VIC costs*, using the same notations as in Sects. 3.1 and 5.3.

Annual fixed costs of VICs in the sphere of generation will naturally amount to a sum of annual fixed costs of individual types of power plants and they do not depend on electricity production:

$$FC_{\text{VIC}} = \sum_{i=1}^{I} FC_i. \tag{5.19}$$

Average fixed costs of VICs depend on its *total* annual electricity production Q_{VIC} (disregarding the output of individual types of power plants):

$$AFC_{\text{VIC}} = FC_{\text{VIC}}/Q_{\text{VIC}} = \frac{1}{Q_{\text{VIC}}} \sum_{i=1}^{I} FC_i. \tag{5.20}$$

It should be emphasized once again that average fixed costs of VICs are determined only by its total annual output disregarding distribution of the output among power plants.

The situation with *variable* costs of VIC is much more complicated. They depend on the annual output of both individual types of power plants (Q_i), and VIC as a whole (Q_{VIC}).

Section 5.3 shows that average variable costs (AVC) virtually of all types of power plants remain fixed throughout the entire range of their *really* possible annual output. For CGPPs they differ for combined and condensing generation of electricity. For further analysis we will *assume* that *average variable costs* of all types of power plants *are fixed* (for HPP they equal zero). Potential violations of this assumption will not radically affect the analysis results. Thus, we assume that

$$AVC_i = const, \; i = 1, \ldots, I, \tag{5.21}$$

where indices "comb" and "cond" will denote combined and condensing generation of a CGPP.

Then *annual* variable costs for individual types of power plants will equal

$$VC_i = AVC_i Q_i, \; i = 1, \ldots, I, \tag{5.22}$$

and average *variable* costs of VIC

$$AVC_{\text{VIC}} = \frac{1}{Q_{\text{VIC}}} \sum_{i=1}^{I} VC_i = \frac{1}{Q_{\text{VIC}}} \sum_{i=1}^{I} AVC_i Q_i. \tag{5.23}$$

This seemingly simple and obvious expression should be commented upon. It characterizes the principal difference between the costs of generation companies and the costs of individual power plants: in the company the *costs* of power plants that belong to it *are averaged.* This particularly manifests itself when VIC has power plants of different types. For example, with HPPs their zero variable costs will decrease the costs of the company as a whole. Besides, for optimization of power system operation according to the minimum variable costs, as was already mentioned, power plants will be loaded or used in the ascending order of these costs. This leads to the fact that average variable costs of VIC will not be constant—their dependence on the annual output will increase (starting with zero, if there are HPPs). This circumstance will be illustrated below by a quantitative example.

The total costs of VIC (TC_{VIC} and ATC_{VIC}) will represent a sum of fixed and variable costs. Using expressions (5.19)–(5.23) we can write for annual total costs of VIC:

$$TC_{\text{VIC}} = \sum_{i=1}^{I} \left[FC_i + AVC_i Q_i \right], \tag{5.24}$$

and for average total costs of VIC

$$ATC_{\text{VIC}} = \frac{1}{Q_{\text{VIC}}} \sum_{i=1}^{I} \left[FC_i + AVC_i Q_i \right]. \tag{5.25}$$

Relationships (5.17)–(5.25) and information from Sect. 5.3 will be used to make a numerical calculation of VIC costs.

5.4.2 *VIC Costs as Applied to the European Section of Russia's UPS*

The main indices of power plants in the European section of UPS (briefly EUPS) were presented in Table 5.1. The EUPS is considered as part of five interconnected power systems: of Center, Northwest, Middle Volga, South and Urals. According to [66] annual electricity consumption in EUPS was assumed (in round figures) to equal 900 billion kWh (900 TWh) and the total installed capacity of power plants—185 GW for the year 2010.

Since the expected shares of gas-fired CPPs with combined-cycle installations (4%) and coal-fired CGPPs (2%) are small (Table 5.1), these types of power plants will be considered in further calculations as gas-fired CPPs with traditional steam turbines (STI) and gas-fired CGPPs, respectively. Thus, consideration will be given to five types of power plants that are presented in Table 5.4 (gas-fired CPPs are with steam turbines).

To show general trends in the change of VIC costs, consideration will be given to two values of electricity consumption (and, hence annual electricity output) in EUPS: $Q_{\mathrm{VIC}}=800$ and 900 TWh. These values will be assumed to include power losses in electric networks of EUPS and represent the total annual electricity generation of the VIC.

The total electricity generation by VIC (800 and 900 TWh) was distributed in an expert way among individual types of power plants, taking into account the above trends in the optimal load distribution among power plants. Thus, annual generation by HPPs, NPPs, and combined generation at CGPPs are assumed equal in both variants. The changes of Q_{VIC} are taken in some proportion of changed output of coal- and gas-fired CPPs and condensing generation at CGPPs. The latter, being

Table 5.4 Installed capacity of power plants and annual electricity production (supply) in EUPS, 2010

Type of power plants	Installed capacity N_i (GW)	$Q_{\mathrm{VIC}}=800$		$Q_{\mathrm{VIC}}=900$	
		Q_i (TWh)	h_i (h/year)	Q_i (TWh)	h_i (h/year)
HPP	21.2	60	3,000	60	3,000
NPP	26.5	180	7,000	180	7,000
Coal-fired CPP	19.3	90	4,500	100	5,000
Gas-fired CPP	48.7	190	4,000	220	4,500
CGPP[a]	69.3	280/280	4,000/4,000	280/340	4,000/4,900
VIC (EUPS)	185.0	800	4,320	900	4,860

[a] Electricity production at CGPP: combined generation in numerator and total generation, including condensing generation, in denominator.

Table 5.5 Annual fixed costs of power plants and generation sphere of VIC in EUPS, 2010

Type of power plant	Installed capacity, N_i (GW)	Specific annual fixed costs[a] ($/kW/year)	Annual fixed costs (*FC*) ($million/year)
HPP	21.2	33.00	700
NPP	26.5	107.25	2,840
Coal-fired CPP	19.3	90.00	1,740
Gas-fired CPP	48.7	66.50	3,240
CGPP	69.3	60.00	4,160
VIC (EUPS)	185.0	–	12,680

[a] See line 6 in Table 5.1 (for CGPP half costs are allocated to heat production).

the most expensive at $Q_{\text{VIC}}=800$ TWh, is not used at all. The distribution of annual output among power plants presented in Table 5.4 is, of course, conventional but realistic enough. This can be seen from the annual number of their installed capacity utilization hours h_i.

The annual *fixed costs* related to generation sphere of VIC are presented in Table 5.5. The costs for individual types of power plants (FC_i) are determined on the basis of their specific values ($/kW/year) that are indicated in Table 5.1. The costs of the entire VIC (FC_{VIC}) represent their sum ($12,680 million a year). For CGPPs, according to the assumption made in Sect. 5.3, 50% of their total fixed costs is spent on electricity production.

As was already mentioned, the annual fixed costs of VIC $FC_{\text{VIC}}=\$12{,}680$ million a year do not depend on its annual electricity output Q_{VIC} and generation by individual types of power plants Q_i. This value will be used to determine average fixed costs AFC_{VIC}.

Variable costs of VIC generation, according to expression (5.23), will depend on both annual electricity output of the entire company Q_{VIC}, and annual outputs of individual types of power plants Q_i. In reality annual output Q_{VIC} at a fixed installed capacity of power plants may vary within relatively narrow limits. For simplification we will assume that in the considered quantitative example it varies in the range from 800 to 900 TWh/year. However, for further comparison with the costs of individual generation companies and illustration of electricity price formation under different market models we will try to conventionally construct the curves of average variable costs of VIC AVC_{VIC} for the entire range of annual output Q_{VIC}, starting with zero.

We will suppose that power plants of different types are used (loaded) in ascending order of their average variable costs AVC_i. Based on this assumption Tables 5.6 and 5.7 present the annual variable costs calculated for individual types of power plants and VIC as a whole at Q_{VIC} equal to 800 and 900 TWh/year. Power plants are presented in ascending order of AVC_i, whose values are taken from Tables 5.1 and 5.3. For CGPPs combined and condensing generation will be considered separately and in combined generation the value of costs assumed is the one for the third method of saving distribution (see clarification to Table 5.3).

It should be noted that the relative efficiency of coal- and gas-fired CPPs (with steam turbines) turned out to be different in total (*ATC*) and variable (*AVC*) costs.

Table 5.6 Annual variable costs of power plants and VIC at $Q_{VIC} = 800$ TWh, 2010

Type of power plant	Power plants			VIC		
	Q_i (TWh)	AVC_i (\$/kWh)	VC_i (\$million)	Q_{VIC} (TWh)	VC_{VIC} (\$million)	AVC_{VIC} (\$/kWh)
HPP	60	0	0	60	0	0
NPP	180	0.0070	1,260	240	1,260	0.0053
CGPP-comb	280	0.0150	4,200	520	5,460	0.0105
Coal-fired CPP	90	0.0158	1,490	610	6,950	0.0114
Gas-fired CPP	190	0.0189	3,590	800	10,540	0.0132
CGPP-cond	–	–	–	–	–	–

Table 5.7 Annual variable costs of power plants and VIC at $Q_{VIC} = 900$ TWh, 2010

Type of power plants	Power plants			VIC		
	Q_i (TWh)	AVC_i (\$/kWh)	VC_i (\$ million)	Q_{VIC} (TWh)	VC_{VIC} (\$million)	AVC_{VIC} (\$/kWh)
HPP	60	0	0	60	0	0
NPP	180	0.0070	1,260	240	1,260	0.0053
CGPP-comb	280	0.0150	4,200	520	5,460	0.0105
Coal-fired CPP	100	0.0158	1,580	620	7,040	0.0114
Gas-fired CPP	220	0.0189	4,160	840	11,200	0.0133
CGPP-cond	60	0.0198	1,190	900	12,390	0.0138

Gas-fired CPPs are more efficient in terms of total costs (Table 5.2) (due to lower specific capital investments), whereas coal-fired CPPs, vice versa, are more efficient in terms of variable costs. Therefore, for the latter the annual number of capacity utilization hours, h, was assumed somewhat higher than for gas-fired CPPs.

The values of annual variable costs of power plants VC_i that are presented in Tables 5.6 and 5.7 are calculated by Eq. (5.22). For VIC, annual output Q_{VIC} is given by a progressive total to 800 or 900 TWh. For example, in the line "CGPP-comb" Q_{VIC} is equal to the sum of the total electricity outputs of HPP, NPP, and combined generation of CGPP. It should be noted that the first three lines (HPP, NPP, and CGPP-comb) in both tables are identical since the output of these power plants is used completely irrespective of whether electricity consumption is 800 or 900 TWh. Similar increase is seen in the annual variable costs of VIC VC_{VIC}. For the same, third line they equal the sum of annual costs of these three types of power plants (for HPP these are equal to zero).

Average variable costs of VIC (AVC_{VIC}) are determined by expression (5.23), i.e., by dividing VC_{VIC} in each line by Q_{VIC}. As is seen due to cost averaging for the entire company, the costs of VIC increase slower than the costs of marginal power plants. In each line the values AVC_{VIC} are lower than AVC_i of a respective type of power plants. In particular, at Q_{VIC}=800 TWh the costs of VIC AVC_{VIC}=0.0132 \$/kWh, whereas the costs of marginal gas-fired CPPs AVC_i=0.0189 \$/kWh (Table 5.6). Similarly, at Q_{VIC}=900 TWh the costs of VIC AVC_{VIC}=0.0138 \$/kWh, while for the marginal CGPP-cond AVC_i=0.0198 \$/kWh or higher by 43% (Table 5.7).

Table 5.8 Average fixed, variable and total short-run (annual) costs of VIC generation, 2010, ¢/kWh

Q_{VIC}, TWh	AFC_{VIC}	AVC_{VIC}	ATC_{VIC}
0	∞	0	∞
60	21.13	0	21.13
240	5.28	0.53	5.81
520	2.43	1.05	3.48
610	2.08	1.14	3.22
620	2.05	1.14	3.19
800	1.59	1.32	2.91
840	1.51	1.33	2.84
900	1.41	1.38	2.79

The data from Tables 5.5–5.7 were used to calculate the average total costs of VIC (Table 5.8). Average fixed costs (AFC_{VIC}) were determined by Eq. (5.20) at FC_{VIC}=$12,680 million/year (Table 5.5), and the values of average variable costs (AVC_{VIC}) were taken from Tables 5.6 and 5.7. The values of all costs of VIC generation are presented in Table 5.8 for the entire range of annual electricity output Q_{VIC} from zero to 900 TWh, though in reality it can vary in the range from 800 to 900 TWh. As was indicated, this was done to have an opportunity to compare the costs of VIC with the costs of individual PGCs that start to operate separately from VIC after transition to the other models of electricity market organization (Models 2–4).

Figure 5.7 shows the curves of average costs for VIC and EUPS that were constructed on the basis of data from Table 5.8. The curve of average fixed costs *AFC* is descending while the curve of variable costs *AVC* is constantly ascending (not U-shaped). The curve of *ATC* reaches minimum at the maximum annual output Q_{VIC}=900 TWh, and in the "working" range of 800–900 TWh the total costs vary insignificantly.

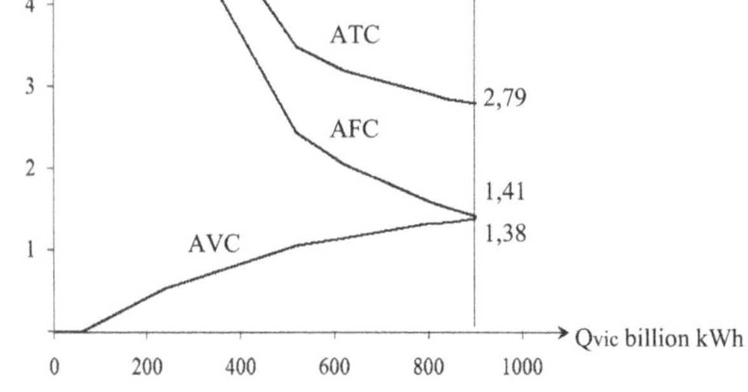

Fig. 5.7 Average fixed, variable, and total short-run costs of VIC, 2010

Let us remind that average annual costs were determined according to the principle of optimal distribution of power plant load on a daily and seasonal basis. Averaging of variable costs of different types of power plants within VIC was also taken into consideration.

The value of average total costs ATC=2.79 ¢/kWh represents weighted average costs of VIC generation at Q_{VIC}=900 TWh. If such a monopoly company is regulated by the State, this value will be included in the tariffs for electricity consumers. Hence, it can be considered as a *wholesale electricity price under regulated monopoly* (Model 1) and compared with the wholesale market prices under the other models of electricity market organization.

It should be noted that under the single-buyer market (Model 2) the sphere of generation is separated from VIC, and several independent PGCs are founded. However, the prices of electricity bought from them are still regulated (i.e., are established at the level of companies' actual costs). If to suppose that the mix of power plants and electricity consumption in EUPS remain the same as those considered for VIC, the *weighted average* price of electricity bought by the Purchasing Agency from a PGC at electricity consumption of 900 TWh/year will be equal to 2.79 ¢/kWh. This can be explained by the fact that dispatching control and optimal distribution of load among power plants will be carried out by Purchasing Agency in the same manner as this was done for VIC.

Thus, for the considered numerical example of European section of UPS of Russia for 2010 the wholesale electricity price will be 2.79 ¢/kWh both under the regulated monopoly (Model 1) and under the single-buyer market (Model 2).

5.4.3 Costs of PGCs and Wholesale Prices in the Competitive Market

Essentially, the costs of PGC that consists of several power plants (probably of different types) will be formed in the same way as the costs of VIC generation. In particular, expressions (5.17)–(5.25) will hold true for PGC. We only have to replace the index "VIC" by the index "PGC" and "i" will denote an individual power plant. The installed capacity, annual electricity output, and annual fixed costs of PGC will, as in the case of VIC, equal the sum of those for individual power plants, whereas average variable costs will be averaged for the PGC as a whole.

Some specific features may reveal themselves under the competitive electricity market (Models 3 and 4) if power plants of PGC are spread throughout the territory of the country (for example like in Russia and China) and appear in different price zones of the wholesale market. In this case the PGC cost characteristics should be determined separately for each zone of the wholesale market.

The costs of PGC (similarly to VIC) will be analyzed for the short run (a year), during which the installed capacities of power plants remain invariable. Here consideration will be given to the specific features of the short-run cost characteristics of power plants that were considered in Sect. 5.3 and certain assumptions will be made to simplify the analysis:

1. The prices, the PGCs or individual power plants offer in *the wholesale electricity market*, are based on their total costs (*TC* or *ATC*) but not on the marginal costs (*MC*) as the theory of microeconomics suggests for "typical" firms (Sect. 2.1). This is explained by practically constant short-run (yearly) average variable costs (*AVC*) and, hence, marginal costs (*MC*) of power plants that are always lower than the average total costs ($MC = AVC < ATC$). With these characteristics, power plants will not pay back fixed costs and will go bankrupt if they follow the "classical" rule according to which the prices offered by firms in the market are based on their marginal costs.
   This circumstance implies that PGC should sell (supply) electricity under long-term contracts, concluded for 1–3 years, with Purchasing Agency (Model 2), with distribution-sales companies (Model 3), and with sales companies or large consumers (Model 4). The prices of the supplied electricity in these contracts should correspond to the average total costs of PGC.
2. In the quantitative analysis of costs the assumption will be made that PGCs *are formed on the basis of one-type power plants* (HPPs, NPPs, etc.). All PGCs with power plants of one type will be considered in total (as one PGC). This makes the analysis of costs more illustrative and enables them to be compared with the costs of VIC that are shown in Fig. 5.7. A more complex picture of costs of individual power plants for the markets in Models 2 and 3 will be given below for the same EUPS on the basis of the actual data for 2003.
3. Annual electricity production by individual PGCs will naturally depend (as in the case of VIC) on load curves of consumers in EPS during a year and on the optimal load distribution among individual types of power plants. For comparability of the analysis results the operation of power plants within a PGC and their annual output are assumed *to remain the same as in the VIC*. The values of average variable and total costs of PGC (*AVC* and *ATC*) will be determined for these annual volumes of production *Q*.

Based on these conditions and assumptions, Table 5.9 presents the PGC costs calculated for the variant of electricity consumption in the European section of UPS of Russia Q_{EUPS}=900 TWh. Conventionally, consideration will be given to five PGCs (for each type of power plants) of the same capacity as done previously.

Table 5.9 Average fixed, variable and total short-run costs of PGCs at $Q_{EUPS} = 900$ TWh, 2010

Name of PGC	N_i (GW)	Q_i (TWh)	FC_i ($ million/year)	AFC_i (¢/kWh)	AVC_i (¢/kWh)	ATC_i (¢/kWh)
HPP	21.2	60	700	1.17	0	1.17
NPP	26.5	180	2,840	1.58	0.7	2.28
CGPP	69.3	340	4,160	1.22	1.59[a]	2.81
Coal-fired CPP	19.3	100	1,740	1.74	1.58	3.32
Gas-fired CPP	48.7	220	3,240	1.47	1.89	3.36

[a] Determined as weighted average of combined and condensing generation of CGPP.

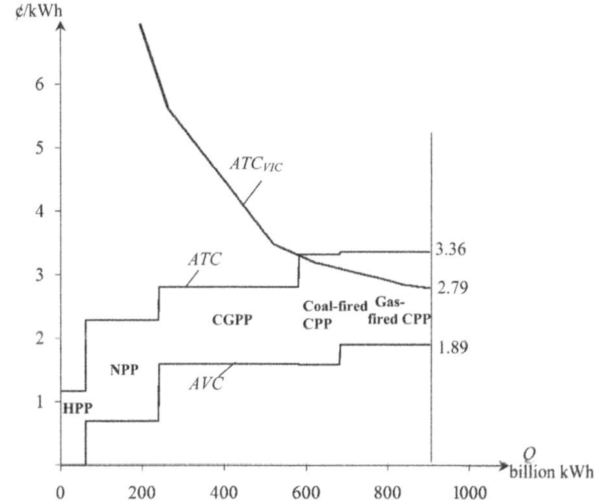

Fig. 5.8 Average variable and total costs of PGCs in EUPS, 2010

Installed capacities (N_i) and annual fixed costs (FC_i) are taken on the basis of data from Table 5.5 and annual electricity production (Q_i) and average variable costs (AVC_i)—on the basis of data from Table 5.7 (for CGPP the combined and condensing generation is considered together). Average fixed costs represent a quotient in division of annual costs by annual output of PGC ($AFC_i = FC_i/Q_i$). The companies are given in ascending order of their average total costs (ATC_i), that are determined as a sum of average fixed and variable costs.

In Fig. 5.8, the values of average variable and total cost of PGC are shown graphically by staged lines. The length of the stages corresponds to the PGC annual electricity output Q_i, and the height to the values of AVC_i and ATC_i at these volumes. The areas below the stages are equal on a corresponding scale to the PGC annual variable and total costs VC_i and TC_i. Note once again that within-year operating conditions and annual output of power plants of all types are taken the same as those in VIC.

For comparison the curve of average total costs of VIC (ATC_{VIC}), shown in Fig. 5.7, is also plotted in Fig. 5.8. It can be seen that at the EUPS annual electricity output Q_{EUPS}=900 TWh average total costs of VIC (2.79 ¢/kWh) are lower than the costs of PGC that has a marginal gas-fired CPP (3.36 ¢/kWh). This circumstance plays an important part in the formation of wholesale prices under different models of electricity market.

Under the single-buyer market (Model 2), the tariffs of electricity supplied by PGCs to the wholesale market are regulated and established at a level of their total costs in the same manner as it is done in VIC. The weighted average tariff of the electricity supplied will be equal to the tariff in the sphere of VIC generation (2.79 ¢/kWh). Hence, the wholesale electricity price under this model will remain the same as under Model 1. Furthermore, it should decrease owing to the effect of competition among electricity producers, provided the tariffs for producers (this was discussed in Sects. 4.1 and 4.2) are regulated appropriately.

The competitive wholesale market with unregulated prices (Model 3) makes the situation change because in the competitive market the equilibrium prices will be formed according to the costs of the most expensive producer (marginal producer), being in demand. This producer in the considered example is PGC with a gas-fired CPP. Its average total costs (3.36 ¢/kWh) exceed the costs of VIC (2.79 ¢/kWh) by 20%. Hence, the transition to the competitive market will increase the wholesale price compared to the regulated market (Models 1 and 2), we can say, "automatically," only because the mechanism of price formation changes. This, as has been mentioned many times, is one of the key flaws of the competitive electricity market.

Potentially, the wholesale electricity price at market organization according to Model 3 may decrease owing to the competition. This, actually, is the main declared target of such a market. However, we should note some circumstances. First, competition among electricity producers can also be implemented in the single-buyer market. Here, as was mentioned in Sect. 4.1, the use of market power by producers is excluded and electricity consumers have a privileged position. Transition from Model 2 to Model 3 in which consumers (distribution-sales companies) start to compete cannot enhance the efficiency of production (decrease costs), and only creates freedom for producers. Second, the decrease in the marginal producer costs to the level of weighted average costs throughout the EPS (the level, existing under price regulation) seems to be rather problematic since it will require large capital investments and time, which will be discussed later.

With the formation of equilibrium price in the wholesale market on the basis of the marginal producer costs (we will call it "marginal" for short) more efficient producers will start to earn additional profit (above normal profit included in the costs). In the theory of microeconomics this extra profit is called producer's surplus.[2]

Similar situation with the wholesale electricity prices will occur under the market organization according to Model 4 in which retail markets are additionally created. Formation of prices in the wholesale market under Models 3 and 4 is practically the same; therefore we can speak of the *competitive* wholesale market, bearing in mind both these models.

The real situation with price formation in the wholesale market while shifting from the regulated (Models 1 and 2) to competitive market (Models 3 and 4) will be more complex, and price rise will be higher than shown in Fig. 5.8, if to take into account that certain power plants of one and the same type differ significantly in their technical and economic indices or if PGCs have power plants of different types. Let us illustrate this by the situation that occurred in the European zone of the Federal Wholesale Electricity and Power Market (FOREM in Russian) in 2003 in Russia. As was mentioned in Sect. 2.3 and will be considered in more detail in Chap. 8, FOREM was organized in the 1990s as a single-buyer market (Model 2). In 2003 there was no free trade sector, and average tariffs of power plants, that were established by the Federal Energy Commission (FEC), reflected very well their annual average total costs (including normal profit).

[2] The definition of the "producer's surplus" is given in Sect. 2.3.

Table 5.10 shows the reported data of 2003 on the basic power plants supplying electricity to the European zone of FOREM. They were placed on the website of FOREM in March 2004.

Power plants are given in ascending order of the average tariff for 2003 (it corresponds to ATC_i) similarly to Table 5.9 and Fig. 5.8. For the picture to be complete, the table presents the fuel used at power plants and interconnected power systems (IPSs) the power plants belong to. The table also shows the calculated total volume of supplies (233.75 TWh) and weighted average tariff for all power plants (41.06 cop/kWh). NPPs which are practically all concentrated in the EUPS are considered on the aggregate (the total volume of supplies and average annual tariff of

Table 5.10 Tariffs and volumes of electricity supplies to FOREM in 2003 by the major power plants in the European part of UPS of Russia. (The table is made up in cooperation with L. Yu. Chudinova and taken from [20])

Power plant	Capacity (MW)	Fuel type	IPS	Average annual tariff (cop/kWh)	Supply volume (billion kWh)
Volzhskaya HPP (Volzhsky city)	2,540	–	Center	11.51	11.29
Kamskaya HPP	480	–	Urals	13.12	1.05
Votkinskaya HPP	1,020	–	Urals	13.84	1.66
Saratovskaya HPP	1,360	–	Middle Volga	15.66	5.78
Volzhskaya HPP named after Lenin	2,300	–	Middle Volga	16.36	9.35
Nizhnegorodskaya HPP	520	–	Center	22.64	1.28
Upper-Volga HPP cascade	450	–	Center	29.10	0.84
Northwestern CGPP	450	Gas	Northwest	35.93	0.98
Kostromskaya CPP	3,600	Gas	Center	35.98	11.11
Permskaya CPP	2,400	Gas	Urals	40.54	11.59
All NPPs	22,200	–	–	42.99	38.75
Nevinnomysskaya CPP	1,270	Gas	South	44.23	4.00
Pechorskaya CPP	1,060	Gas	Northwest	45.14	2.25
Kirishskaya CPP	2,100	Gas	Northwest	52.35	2.65
Stavropolskaya CPP	2,400	Gas/fuel oil	South	53.06	6.27
Konakovskaya CPP	2,400	Gas	Center	54.02	5.77
Ryazanskaya CPP	2,720	Gas/coal/fuel oil	Center	56.81	6.21
Troitskaya CPP	2,060	Coal	Urals	57.83	4.45
Pskovskaya CPP	430	Gas	Northwest	59.84	1.35
Novocherkasskaya CPP	2,400	coal/fuel oil/gas	South	65.73	5.69
Cherepetskaya CPP	1,500	coal/fuel oil	Center	96.13	1.43
			Total	41.06[a]	233.75

[a] Weighted average tariff of all power plants.

NPP) which is of no importance in this case. Only one CGPP entered FOREM while the rest of them (i.e., almost all) supplied electricity to the local power systems (AO-energo) representing regulated monopolies at that time.

As is seen from Table 5.10, HPPs have the lowest tariffs (11.51–29.10 cop/kWh) and costs. The HPPs are followed by a group of three gas-fired thermal power plants situated in various IPSs (in 2003 unlike the forecasts for 2010 the internal regulated natural gas prices in Russia were lower than the market coal prices). Their tariffs were lower than the average tariff for NPPs. The tariffs of the other CPPs were higher than those for NPPs. The Novocherkasskaya and Cherepetskaya CPPs that burn mainly coal and fuel oil have the highest tariffs.

The FOREM website also contained the information on the average annual tariff for buyers in the interconnected power systems that belonged to the EUPS in 2003. It equals nearly 45.5 cop/kWh and exceeded annual average tariff for power plants (41.06 cop/kWh) by the value of all-system costs of RAO "EES Rossii," including the network access fee.

Figure 5.9 shows a staged line of tariffs (similar to Fig. 5.8) for HPPs, CPPs of the first group, NPPs and CPPs of the second group. The line was constructed on the basis of data from Table 5.10. The stages of this line characterize the amount of electricity supplied in 2003 at a respective tariff. The number of stages (except for NPP) is naturally much larger than in Fig. 5.8. This line can be considered as an aggregate supply curve of producers *S* in the competitive market.

Dashed lines are used to show the annual average tariff for power plants and average annual tariff for consumers. The latter, as was already mentioned, includes the costs of power plants (i.e., the sphere of generation in the EUPS) and system costs, included in the wholesale electricity prices. At this tariff 233.75 TWh was sold,

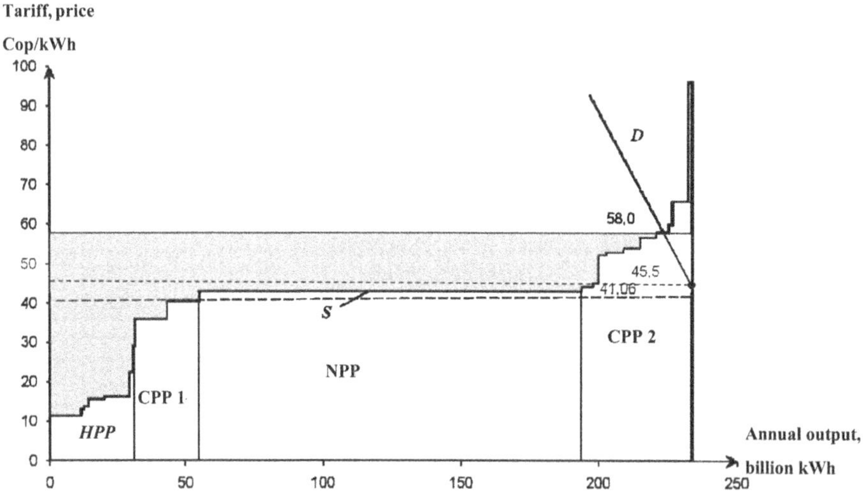

Fig. 5.9 The curve of total costs of power plants, marginal price, and producer's surplus in the European part of UPS of Russia, 2003

which can be considered as a solvent demand of consumers. The line D standing for the demand curve of consumers has been drawn through this point with somewhat conventional tilt.

If to imagine that electricity prices in FOREM in 2003 were not regulated (if there were the competitive wholesale market), then there would be an equilibrium (marginal) price equal to nearly 58 cop/kWh. Then all power plants would sell electricity at this price and gain additional profit (producer's surplus) equal to the dashed area. Quantitatively, it makes up about 40 billion rubles a year or approximately 35% of actual annual costs of power plants (the areas under the staged line *S*). In fact, the marginal price would be even somewhat higher since under the competitive market the costs of power plants would rise additionally by the amount to be paid to System Operator, Trading System Administrator, for network access, etc.

It can be seen that the producers' surplus resulting from price rise is in no way connected to the enhancement of electricity production efficiency and is not the achievement of producers.

It is conditioned by the mechanism of price formation in the competitive (non-regulated) markets and EPS features, among which the most important are the following:

1. A special shape of curves of the average costs of power plants, which is shown in Sect. 5.3. On the one hand, it excludes (or makes erroneous) organization of the *spot* electricity market in real time since in the spot market the fixed costs of power plants are not recovered. On the other hand, power plants should enter the really *short-run* (annual) market with their prices based on their average *total* costs (*ATC*) rather than with the marginal costs (*MC*) that are always lower than the total ones. Note that the "typical" firms considered in the microeconomics have short-run cost curves of U shape, and the firms enter the market with supply curves (*S*) and volumes of goods at which *MC* exceed *ATC*, which makes it possible to fully recover all the costs and even gain additional (economic) profit.
2. The presence (feasibility of construction) of different types of power plants in EPS (Sect. 2.3) which, with EPS operation optimized with regard to the *instantaneous* (hourly) cost characteristics, will be economically efficient in various zones of daily load curves (basic, semi-peak, and peak). However, *annual* costs of these power plants differ considerably and in the short-run competitive wholesale electricity market the equilibrium prices are formed by *annual* costs of *marginal* power plants (i.e., those with the highest costs). Hence, unlike the market for other goods, the most important for the competitive wholesale electricity market is the *differences* in the *total* costs of various types of power plants rather than in the marginal costs of various firms. This circumstance is often overlooked when electricity markets are analyzed.
3. The existence of a physical barrier to new producers' entry into the market. Only operating producers (with fixed installed capacities) participate in the short-run electricity market and it is absolutely impossible for new producers to affect the prices. NPPs can appear only in the long-run market which will be considered in Chap. 6.

These and some other properties of EPS lead to the fact that transition to the competitive wholesale market (when prices are no longer regulated) which occurs at invariable capacities of power plants will make electricity prices rise to the marginal ones which correspond to the total costs of the marginal power plant. A paradoxical situation occurs: the competitive market is created with a hope that competition will enhance the efficiency of electricity production and decrease its prices; meanwhile as soon as this market is introduced the prices will rise to the level of the most expensive marginal producer.

It is often stated that in the future the prices will drop as a result of cost decrease due to competition. However, the author did not find any quantitative substantiations of this statement: what power plants should be replaced by new, more efficient ones, what investment is required, how much time it will take, what will be the dynamics of price decrease, etc. If anyone had ever tried to make such a substantiation, they apparently would have not dared publish it because it would definitely have not been in favor of a competitive market. In order to decrease marginal prices by 30% (this is an approximate amount by which they will exceed the weighted average tariffs at transition to the competitive market) it will be necessary to replace more than half power plants (with appropriate investments) and it will take several decades to do that. The most important point here is that under regulated electricity markets (Models 1 and 2) similar decrease can be achieved for weighted average tariffs without increasing prices to the marginal ones.

5.4.4 About Market of Long-Term Contracts

As has already been noted, the normal competitive market along with the direct trade in some goods should generate "price signals" concerning the production volumes in the short run and market expansion (or shrinkage) in the long run. This happens in the markets for most (or many) goods including spot markets (trade on the spot). With no price signals (or their distortion) market will be inadequate, i.e., an obviously imperfect market which does not perform important functions, spontaneous, etc. For the power industry it would be inadmissible.

Real time spot electricity markets which, according to the plan of developers of the competitive market concept, were intended for these purposes, turned out to be inappropriate. The reasons (there are several of them) were analyzed in Sect. 5.2. A significant fact is that Great Britain, whose market is often taken as a model to follow, refused from the "day ahead" market even in 2001 when the market conception was changed (to NETA).

Due to special conditions of electricity production and EPS properties, electricity trade should be based on the *long-term contracts* (for the period of 1–3 years) in which the prices reflect *total* costs (including fixed ones). This refers to both the wholesale electricity markets with regulated (Models 1 and 2) and unregulated prices (Models 3 and 4). In doing so, it is necessary to provide:

- On the one hand, *current* electricity consumption (and its reliability) in the short run from operating power plants (with fixed capacities).
- On the other hand, *perspective* electricity consumption in the long run with account taken of EPS expansion and construction of new power plants. In this case the costs of PGCs or individual (new) power plants will include a certain investment component (along with pure operating costs).

With regulated prices (tariffs) of VICs (Model 1) and in the single-buyer market (Model 2) the price signals are not required. They are replaced by centralized planning of annual operating conditions and expansion of EPS, to be carried out by a monopoly company or Purchasing Agency and to be coordinated afterwards with the regulatory body. The tariffs for VIC generation or individual power plants are established on the basis of actual electricity production costs (taking into account the rate of profit, operation of power plants, investments etc.) When necessary (particularly under the single-buyer market) the values of electricity tariffs and supply volumes are established by the long-term contracts. As was noted in Sects. 4.1 and 4.2, market organization according to Models 1 and 2 provides optimal operation and expansion of EPS and tariffs for consumers are established at the level of average (or weighted average) costs of power companies throughout the EPS as a whole. The degree of "optimality" depends to a certain extent on the quality (perfection of principles, methods, and procedures) of the state regulation. When the state regulation is adequate, the problems with control of EPS expansion and operation do not arise.

Much more difficult situation occurs under organization of the competitive wholesale market (Models 3 and 4). In this case there should be a competitive market of long-term contracts, as a rule, bilateral (between certain producers and buyers of electricity). Participants in this market should include both the existing producers providing current electricity consumption and new ones aimed to cover the expected increase in consumption. Let us consider how this may be organized.

The only example of such a market, known to the author, is *NETA* market (*New Electricity Trading Arrangements*) in Great Britain. It was introduced in 2001 and transformed into *BETTA* market (*British Electricity Trading and Transmission Arrangements*) in 2005. The concepts of *NETA* and *BETTA* suggest organization of a *forward market of standardized long-term contracts* for the period of several years [69]. This can be supposed to be the competitive market of long-term contracts we are discussing. The participants in this market are producers and buyers, the prices and volumes offered by producers are known (the prices are based on the total short-run or long-run costs), the bids of buyers are known, and therefore, the equilibrium market price will be formed according to supply and demand. Such market can generate the required price signals.

Unfortunately, as indicated in [69] this segment of BETTA market has not been organized yet (the reasons are not discussed). Hence, there is *no world's experience in creating the true competitive market of long-term contracts* which is necessary in power industry. This can be explained apparently by the fact that, on the one hand, the need for this market is not understood everywhere and the trend towards spot electricity markets in real time is persistent, while on the other hand, the countries that are aware of the need for this market (in particular in Great Britain) encounter certain difficulties in organizing it.

Long-term bilateral contracts are concluded in Great Britain at out-of-exchange sites, i.e., in the form of individual transactions. Electricity prices in these contracts are, as a rule, confidential (this is pointed out in [4], for example). No price signals are formed, which confirms once again the electricity market imperfection.

It should be noted that the market of long-term contracts cannot be considered as a spot market since the trade (exchange of a commodity for money) does not occur "on the spot." Long-term contracts are concluded for future electricity deliveries and stipulate conditions and terms of delivery, and payment. Therefore, taking into account inappropriateness of electricity markets in real time we can speak of the impossibility to organize spot electricity markets.

Box 13 Formation of Short-Run Costs of Generating Companies and Wholesale Electricity Market Prices

1. In the generation sphere of VICs and PGCs the short-run costs of their power plants are *averaged*. For VIC the costs are averaged for all power plants in EPS on its territory, while for PGCs, which are supposed to be separated from VIC, the costs are averaged for power plants that belong to a certain PGC only.
2. The prices of the short-run wholesale market basically depend on whether they are regulated (Models 1 and 2) or not (Models 3 and 4).
3. Under regulation the wholesale prices are set at the level of weighted short-run (yearly) average total costs (*ATC*) in the sphere of EPS generation. If the mix of power plants is constant, the prices will approximately be the same in Models 1 and 2 (in Model 2 the prices can be reduced owing to the effect of competition among electricity producers).
4. Unregulated prices in the competitive wholesale market (in Models 3 and 4) will lead to the formation of equilibrium (marginal) prices at the level of average total costs of the least efficient PGC (or rather a marginal power plant) in EPS. In the concrete example of the European part of UPS of Russia the equilibrium prices will be about 30% above the regulated prices. Further price decrease by such a value due to competition seems unreal.
5. When the competitive wholesale market prices rise to the marginal ones, more efficient PGCs (and power plants) will start to receive extra profit, i.e., "producer's surplus."
6. This extra profit is not a result of producers' efforts or skills. It is gained owing to the properties of unregulated competitive markets. When prices are regulated (in Models 1 and 2), this surplus is withdrawn from producers, which results in lower prices for electricity consumers.

7. As far as the author knows, the *competitive* market of long-term contracts, which *is required in power industry*, is foreseen only in Great Britain in the concepts of *NETA* and *BETTA* markets. A *forward market* (*exchange*) *of standardized long-term contracts* for the period of several years had to be one of the BETTA segments. However, for some reasons this segment of the market has not been created so far. Long-term bilateral contracts are concluded beyond exchange sites, in the form of individual transactions with *confidential* prices.

Chapter 6
EPS Expansion Under Different Market Models

This chapter gives consideration to the issues and problems related to EPS development. Special attention is focused on the expansion of generation capacities and intersystem electric ties (ISETs), and on wholesale electricity markets in which power plants and ISETs participate. Development of intrasystem (transportation and distribution) electric networks which remain monopoly and regulated spheres in all market organization models is not considered.

Analyses of the mechanisms for financing new power plants (Sect. 6.1), the required investment component of electricity tariffs or prices under different market models (Sect. 6.2), long-run costs in the sphere of EPS generation (Sect. 6.3), and prices of competitive wholesale market in the long run (Sect. 6.4) are given. Additionally, Sect. 6.5 shows the specific features of substantiating the efficiency of intersystem and interstate electric ties in the electricity markets with regulated (Models 1 and 2) and unregulated (Models 3 and 4) prices.

6.1 Financing Mechanisms for Construction of Power Plants

The main features of financing the expansion of generation capacities in EPS under different market organization models have already been discussed in Chap. 4:

- In regulated monopoly (Model 1) financing is made by including the required investments into the *investment component of tariffs for consumers*. There can be two ways: (a) direct inclusion of investments in tariffs *during* the power plant *construction period* (this way will be called "self-financing" since the monopoly company (VIC) itself finances the construction by the revenues earned from selling electricity at established tariffs) and (b) construction at the expense of bank credits (in this case the investment component includes *credit repayment*). In both cases the investments or credit repayments are allocated among all consumers, i.e., divided by all volumes of electricity sold by the vertically integrated company (VIC).

L. S. Belyaev, *Electricity Market Reforms*,

DOI 10.1007/978-1-4419-5612-5_6,

- In the single-buyer market (Model 2) power plants are constructed by *private investors*—future owners of the power plants (power-generating companies (PGCs) or new power producers (NPPs)). The company Purchasing Agency concludes (on the basis of tenders) long-term regulated contracts (for 10–15 years) which stipulate the prices ensuring investment recovery during the contract period. Higher electricity prices in such contracts and the prices of operating producers are averaged and an *average weighted* tariff is established for consumers. The tariff provides the *recovery of investments* in the new power plants, i.e., contains, in fact, the same investment component as in the regulated monopoly which develops at the expense of credits. In general, investment mechanism at market organization according to Model 2 turns out to be identical to the second way under Model 1, only instead of *credit* repayment there will be *investment* repayment, which is still distributed among all volumes of electricity supplied to consumers in this EPS.
- Under competitive wholesale market (Models 3 and 4) new power plants will also be constructed by *private investors*, though under *unregulated* prices. This introduces essential distinctions in the *mechanism of investment recovery* as compared to Model 2. Investment recovery is *no longer guaranteed* to the investor. Whether or not the investments are recouped will depend on prices to be set in the wholesale electricity market. These prices are highly volatile and uncertain, therefore the investor will be at risk of not recovering or even losing the capital invested. Besides, as will be discussed below, due to the absence of regulation the investment in any new power plant should be recouped only at the expense of electricity generated by this very plant (but not at the expense of all electricity bought by consumers in the EPS).

The indicated conditions of financing the construction of new power plants under different models of wholesale market organization have both common features and principal differences. Therefore, it is necessary to analyze (take into account):

- The conditions for repayment of credits or recovery of private investments, i.e., whether they are guaranteed or nonguaranteed (risk-bearing).
- Electricity volumes among which the investments are apportioned or at the expense of which they are recovered. This can be either all electricity consumed (purchased) in power systems participating in the wholesale market (Models 1 and 2) or electricity produced (supplied) only by one power plant in which the capital is invested (Models 3 and 4).

The combination of these factors will be called a *financing mechanism*. A certain model (formula) will correspond to each of these mechanisms, which is used to calculate a quantitative value of the investment component of electricity prices and tariffs. These formulas will be given in the next section. Here we will dwell on the types and essence of various financing mechanisms.

Formation of financing mechanisms is shown in Table 6.1. The two above-mentioned ways are considered for the regulated monopoly: "self-financing" and "crediting." Models 3 and 4 are written together since the prices in the competitive

Table 6.1 Factors and conditions that determine the financing mechanisms for power plant construction

Model of electricity market organization	Financing source	Conditions of credit and investment repayment	Volumes of electricity to recover investments	Number of financing mechanism
Model 1	(1) VIC funds	–	Of the entire VIC	1
	(2) Bank credits	Guaranteed	Of the entire VIC	2
Model 2	Private investor	Guaranteed	Of the entire EPS	2
Model 3	The same	Risk-bearing	Of one power plant	3

wholesale market are only important for financing power plants (sometimes Model 4 will be skipped).

It is seen that the self-financing mechanism of regulated monopoly (Mechanism 1) differs essentially from the other mechanisms since the investments are included directly in the investment component of consumer tariff (during the power plant construction period). Bank credits or private investments are not used. At the same time this mechanism has a feature common for all electricity markets with regulated prices (Models 1 and 2), i.e., the investment component is included in the tariffs for *all consumers* supplied by a monopoly VIC or by all PGCs and NPPs on the territory of this EPS. Thus, the investments in power plants that are constructed in some year are distributed among all volumes of electricity consumed in EPS this year.

In fact, the financing mechanisms for construction of power plants under regulated monopoly using credits and single-buyer market are identical. During the power plant construction period consumers do not bear any costs related to the construction (the power plant is constructed for them "free of charge"). However, afterwards consumers pay the investment component, which includes the repayment of credits or recovery of private investments and a respective interest. In both cases this repayment is guaranteed and distributed, as in Mechanism 1, among all volumes of electricity consumed in this EPS. Therefore, such mechanisms will be considered as Mechanism 2.

Financing of power plant construction under conditions of a competitive market differs radically from the previously mentioned ones, though similar to Model 2 in terms of financing source. As was mentioned above, investment recovery is no longer guaranteed to the investor. This will result in an increasing interest on capital, which will encourage the investor to make the investments. Importantly, now the investments should be recouped only at the expense of power generated by this one power plant being under construction (this circumstance will be explained below). This combination of conditions and factors of power plant construction is identified as Mechanism 3.

Let us explain the term (or notion) "private investor." It will be applied in general to all potential investors when the sphere of EPS generation is split into several financially independent companies (Models 2–4) versus monopoly companies regulated by the State. Private investors can be power-generating companies (PGCs), which separated from VICs after restructuring, or NPPs, which in the beginning construct only one power plant. The investors for NPPs can be nonenergy and foreign companies. Sometimes PGCs can be state-owned, but possess the necessary economic independence.

We will assume that private investors:

- Have their own capital (even if it is partially borrowed) and make investment decisions themselves.
- Can invest in a new power plant and other alternative projects, including those in fields other than the power industry.
- Will own the constructed power plant and operate it.
- Are completely aware of both benefits and negative consequences (risks) related to the construction of the power plant.

The situation encountered by private investors in the single-buyer market differs considerably from that in the competitive market.

In the single-buyer market it is important for an investor to win the tender for the construction of a new power plant with an acceptable price of electricity to be supplied. This price should provide recovery of investments in some period T_R with an acceptable interest on capital σ (and covers all the operating costs). Undoubtedly, the indices T_R and σ for power plant investments should be better than in the other alternative capital investments (otherwise the investor will refuse to construct the power plant). The investor will also take into account the *guaranteed* recovery of the investment in electricity which will not necessarily be the case in the alternative projects. Therefore, he may decrease the desired interest on capital σ and increase the period of its recovery T_R. Tentatively we can suppose that the interest on capital under its guaranteed recovery will make up $\sigma = 0.03$–0.08.

After winning the tender and concluding a long-term contract with the Purchasing Agency for electricity supply at an acceptable price during period T_R, the investor will not be interested in the wholesale electricity market prices. Under any prices (depending on supplies of other producers), after constructing the power plant the investor will sell his electricity at the price stipulated by the contract. This price will be higher than the average weighted wholesale price because of the new power plant investments to be recovered. However, this difference will be included in the investment components of tariffs and distributed among *all* volumes of electricity consumed in the EPS.

The situation encountered by the private investor in the competitive market will be principally different:

1. Investment *risk* will be *borne by the investor* only.
2. The financial efficiency of *each project* for the construction of a new power plant will be assessed *individually*.
3. *Investments* in a certain power plant *should be recovered* at the expense of electricity produced by *this one* power plant only.
4. *The costs of existing* power plants (and, probably, the wholesale market prices) will *be lower* than the prices to be offered by similar *new* power plants.

In the competitive market the *risk* due to erroneous decisions leading to long payback periods or even loss of investments is not borne by consumers as was the case in the regulated single-buyer market. The investor should make his own decision and deal with its consequences. Recovery of investments depends first of all on future electricity prices. The investor needs the forecasts of the wholesale market prices for the coming 15–20 years, which include the power plant design and construction period as well as the investment payback period. These forecasts will naturally be highly uncertain, and this will increase the investor's risk and raise interest on capital σ, at which the investor will undertake to construct a new power plant. According to the available estimates [70, 71] this increase makes up 7–9% ($\Delta\sigma = 0.07$–0.09). Thus, the interest σ in the competitive market can be expected to increase up to 0.12–0.20.

It is obvious that financial efficiency of power plant construction should be *estimated individually* by each NPP, i.e., the future owner of his first power plant should be sure that this investment is efficient. In the case of the *existing* PGCs the inevitability of individual estimation of the construction efficiency of each new power plant is less obvious. There are opinions that in the competitive market independent PGCs will construct (and recoup) *new* power plants at the expense of funds received from the selling power of *all operating* power plants they own, similar to what the regulated monopoly companies did. However, this is not so if to analyze more thoroughly concerns and possibilities of independent PGCs.

First suppose that some PGC starts to include the investment component in the electricity prices it offers in the wholesale market. Then, all other conditions being equal (at the same mixes of power plant) it will lose to other PGCs which do not do that. This company will lose market and will not be able to operate normally, as the investment component has to be included in the prices during several years of the new power plant construction. Besides, as was mentioned in Sect. 3.2 when considering oligopoly, the existing PGCs are not interested in the emergence of new power plants in the market at all, since this will increase supply and decrease prices. They benefit from power shortage. Therefore, the existing PGCs practically rule out such a way of constructing new power plants.

Hence, the existing PGCs will be able to construct new power plants only if they *accumulate* capital beforehand. The accumulation is possible at the expense of:

- Depreciation charges
- Producer's surplus
- Monopoly profit, if there is capacity shortage in the wholesale market
- Activity not related to the electricity production

Now imagine that the existing PGC has accumulated capital and decides how to use it. The following circumstances should be taken into account:

- The PGC, as any private company, will seek to invest its capital in the most profitable way.
- It makes no difference to the PGC what projects to invest in, and it will certainly not construct a new power plant if there are more profitable variants of investments; this follows from the possibility for independent PGCs to invest capital into any sector of the economy.

- The PGC will uniformly estimate and then compare the financial efficiency of constructing a new power plant and alternative projects: business plans will be drawn up; financial flows and profiles will be calculated; the investment payback periods, net present value, internal rate of return, etc., will be determined.

Thus, the existing PGCs will estimate the financial efficiency of a new power plant only as a variant of the investment of the capital formed. This variant will naturally be estimated *individually* for a certain project of a new power plant in comparison with alternative projects. And this estimation, in fact, is similar to the estimation to be made by an NPP.

Estimation of the financial efficiency of investment projects calls for rather cumbersome calculations [72–74]. It is necessary to take into account investments, operating costs, revenues from selling products, depreciation, taxes, inflation, etc. As a rule, the calculations are made for the long run, which covers the period of construction and service of the facility, with the investor's revenues and costs discounted and reduced to a certain time instant (for example to the year of project launch). The calculation results are used to determine the indices of the project efficiency and profitability (there are several of them) which are then employed by the investor to take either a positive or a negative decision on investment in the considered project.

The main values which affect the project efficiency are the volume of investments, annual operating costs, and annual revenue from selling commodities produced by the constructed facility. This revenue should compensate for the operating costs and recover the investments during some time. For the project of a new power plant (as an alternative of PGC capital investment) the revenue will be determined by the amount of electricity generated by this power plant, and the price at which this electricity is sold. Hence, the investments should be paid back by selling the electricity generated by this *one power plant only* (during the repayment period T_R).

This increases the investment component of electricity price as compared to the investment component in tariffs under regulated markets, where investments are distributed among the outputs of all power plants in EPS. For Russia the *price increase* at transition from self-financing monopoly to private investments under competitive market was estimated at 2–3 ¢/kWh [17, 54–56].

The indicated distinctions of the new power plant investment lead to the situation where the costs of *operating* power plants that determine the wholesale market prices under competitive market will knowingly be lower than the prices required to attract investments in *similar new* power plants. This will happen both at transition from regulated to competitive market and further when this market will operate for a sufficiently long period of time [19].

For further analysis it is important to note once again that in Mechanism 2 (Table 6.1) where power plants are constructed at the expense of bank credits or private investors with *guaranteed* payback of credits or investments, the interest on capital σ will be lower than for the construction of power plants under a competitive market (Mechanism 3). The approximate values of the interests have already been indicated earlier: $\sigma_2 = 0.03$–0.08 and $\sigma_3 = 0.12$–0.20 (indices "2" and "3" will further denote Mechanisms 2 and 3). Thus, there will always be inequality:

$$\sigma_2 < \sigma_3. \tag{6.1}$$

It is also necessary to take into account the EPS property that was considered in Sect. 2.3—the *long service life of power plants* T_L. These service lives, as a rule, make up 30 years and more. Such long periods obviously exceed the periods of investment recovery T_R, which can be acceptable to private investors (10–15 years and even shorter):

$$T_R < T_L. \tag{6.2}$$

It follows that during most of its service life (after payback of credits and investments) a power plant will bear only operating costs. Under the regulated markets (Mechanism 2) this will be taken into consideration by excluding the recovered amounts from the investment components of consumer tariffs. Under the competitive market after investment recovery power plants will have an extra profit that can be called a monopoly profit since it occurs owing to excess of price over costs.

Thus, it is possible to distinguish three main financing mechanisms for construction of power plants:

- *Mechanism 1—Self-financing* expansion of generation capacities under regulated monopoly (Model 1)
- *Mechanism 2*—Construction of power plants *at the expense of credits* under regulated monopoly (Model 1) or by *private investors* under the single-buyer market (Model 2)
- *Mechanism 3*—Construction of power plants by *private investors* under the competitive wholesale market (Models 3 and 4)

In Mechanism 1 investments are included in tariffs with a required lead time, and electricity consumers pay these investments during the period of power plant construction. After putting the power plant into operation consumers do not bear any costs related to its construction.

In Mechanisms 2 and 3 the picture is somewhat different: electricity consumers do not pay the investments during the period of power plant construction, but afterwards, during the repayment period T_R, they will pay them with a certain interest σ. Payback of credits or private investments gets protracted; however, the total amount of payment rises because of the interest σ.

For greater clarity the three ways of financing the expansion of EPS generation capacities will be considered in the "pure" form, i.e., without dwelling on their possible combinations.

The prices in the competitive market (Mechanism 3) are naturally formed on the basis of demand and supply and are uncertain for the future. The situations are possible when they will turn out low and the investments in the new power plants will not be recovered at all. Therefore, in further analysis we will consider the *prices required* to recover the private investments in the period T_R set by the investor. Based on this period (and interest on capital σ, also set by the investor) it is possible to determine the required value of the investment component and the required electricity price by adding this component to the operating costs which will depend on technical and economic indices of the power plant.

Let us consider now the *procedure of including capital investments* in electricity tariffs and prices as applied to one new power plant. This is important because under any financing mechanism the *investments in the new power plants are paid by consumers*, but sometimes they may also pay the extra profit formed.

This procedure is shown in Fig. 6.1 for the construction period T_C and the service life of the power plant T_L for three considered financing mechanisms. The expenditures E stand for the annual costs (million \$/year) for one and the same power plant.

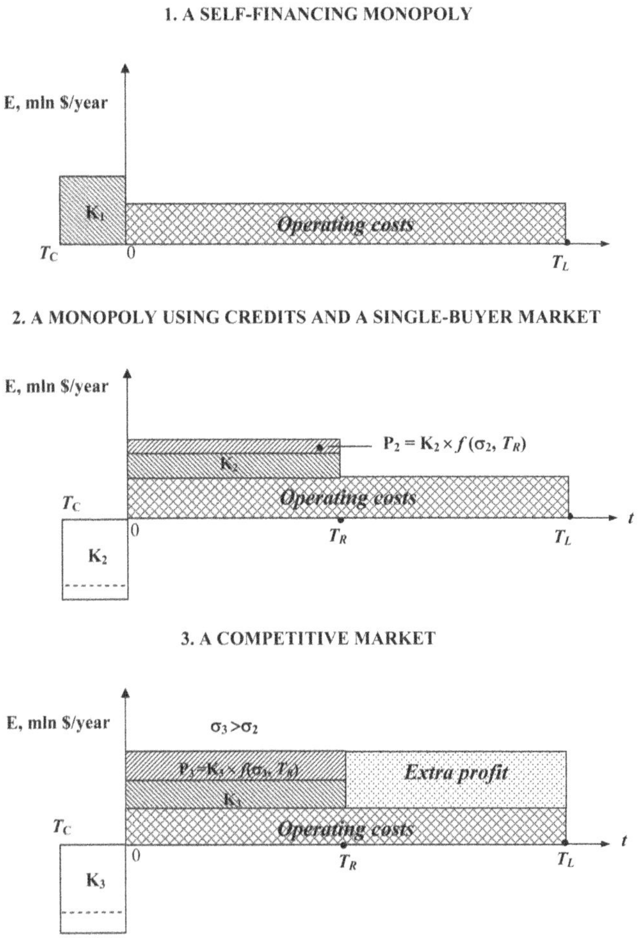

Fig. 6.1 Expenditures for new power plant construction and operation that are included in the electricity price (tariff)

For a *self-financing monopoly* the area K_1 represents total capital investments in power plants (their distribution by year of construction period T_C is supposed to be even). They are included in consumer tariffs during the construction period. During the period of operation, tariffs include only operating costs. The figure shows them to be constant throughout the entire service life of the power plant T_L.

It should be reminded that here the investments and costs of only one power plant are discussed. As to the tariff for the entire monopoly company, it will be based, first, on the average costs of all operating power plants. Second, this tariff will include an investment component that embraces the annual volume of investments of all simultaneously constructed power plants, but this volume, as was mentioned many times, is distributed among the volumes of electricity produced by all operating power plants in EPS.

In *the monopoly developing at the expense of credits and in the single-buyer market*, power plants are constructed free of charge for consumers (the area K_2 is not dashed), but the capital investments K_2 increase as compared to K_1 due to the interest added during construction period. As operation starts, consumer tariffs (over the repayment period T_R) include this increased value of K_2, and again with the interest σ_2 added. The value Π_2 represents profit of a creditor or investor, which depends on the interest σ_2 and the repayment period of credits or investments T_R.

After the credits or investments are repaid the tariff will include, as in the previous case, only "net" operating costs. This happens because the regulatory body includes the repayment of credits or investments (with interests) in the tariffs only when they are really paid. This is a major difference from the competitive market.

Tariffs for the entire monopoly company or Purchasing Agency are formed as in the case of a self-financing monopoly, i.e., on the basis of average operating costs of operating power plants with an investment component added. The latter includes repayment of credits or investments of all new power plants which have not been paid back yet, and these payments are distributed among all volumes of electricity supplied to consumers in EPS. As is shown in the next section, the value of the investment component for the monopoly using credits or Purchasing Agency may be higher or lower than for the self-financing monopoly, depending on the relationship between the interest on capital σ and development pace λ.

For a *competitive market* Fig. 6.1 is constructed on the assumption that the wholesale electricity market prices are high enough to attract private investors (otherwise power plant would not be constructed). In this case the interest on capital σ_3 is higher than in the previous case due to higher financial risk. Therefore, the sum of capital investments to be recovered and the profit of investors additionally rise ($K_3 > K_2$ and $\Pi_3 > \Pi_2$).

If electricity price after investment recovery continues to be above operating costs (including normal profit) the power plant owner will reap additional profit. And consumer buying electricity will pay this extra profit along with the really required costs related to the EPS development. Hence, under the free market, electricity prices can include additional profit of power plants that have already recovered their investments.

In Fig. 6.1 the extra profit is shown on the assumption that electricity price in the wholesale market remains constant during the entire service life of the power plant T_L. In fact, the competitive market is characterized by price fluctuations depending on shortage or surplus of generation capacities. If to imagine that putting into operation the power plant shown in the figure results in capacity surplus and price drop in the wholesale market, the investments in this power plant will not be recovered. Similarly, the recently constructed power plants will also cease to recover their investments. Moreover, the new power plants, the need for which will arise in 5–10 years, will not be constructed either. Therefore, it is logical to suppose that the investor who constructed the considered power plant was sure that the high price level he needed would be maintained for a long time. This could happen only at permanently high wholesale market prices to be formed at the moment of decision making on power plant construction. Hence, permanently high prices are necessary to provide expansion of generation capacities under the competitive market. The prices can vary in some range, but on the whole their level will be such that power plants that have recovered their investments will gain the increased (i.e., monopoly) profit.

Box 14 Financing Mechanisms for Construction of Power Plants

1. We can single out three principal financing mechanisms:

   Mechanism 1—*Self-financing* of generation capacity development under the regulated monopoly
   Mechanism 2—Construction of power plants *at the expense of credits* in the regulated monopoly and by *private investors* in the single-buyer market
   Mechanism 3—Construction of power plants by *private investors* in the competitive market

2. In Mechanism 1 the investments in power plants are included in the investment component of consumer tariffs *directly in the course of their construction* and then consumers cover only operating costs. In the other two mechanisms power plants are constructed "free of charge" for consumers. However, after they are put into operation, *consumers repay credits or private investments* during the repayment period T_R with the interest on capital σ.
3. Mechanism 2 *guarantees repayment* of credits or private investments, hence the interest on capital σ will be lower and the repayment period T_R can be longer than in Mechanism 3 characterized by *high-risk* investments.
4. In Mechanisms 1 and 2 the investments in new power plants are distributed among all volumes of *electricity* supplied to consumers of the considered VIC or EPS, which makes the investment component of tariffs much lower than in Mechanism 3, where the investments should be repaid owing to electricity produced *only by one* (this) new *power plant*.

5. In Mechanism 3 producers have an opportunity to gain the monopoly profit (superprofit) that is paid by electricity consumers.

6.2 Models of Price Formation and Their Analysis[1]

The section considers the equations for the investment component r and prices or tariffs p for the three considered financing mechanisms for expansion of generation capacities in EPS. They are analyzed and compared. For all the three cases the electricity price (tariff) is represented in a uniform manner

$$p = r + c, \tag{6.3}$$

where c is electricity generation costs which are supposed to be identical under all market organization models. This provides comparability of tariffs and market prices.

Some simplifications and assumptions were made when deriving the equations to make the analysis and demonstration of major regularities more convenient:

- Consideration is given to *power plants of one and the same type* with identical and invariable technical and economic indices for both operating and new power plants. The entire EPS is supposed to consist of such power plants only. For the competitive market the possibility of costs decrease owing to competition will be pointed out.
- The power plant *construction period is not taken into account*. The capital investments in some year t are supposed to provide the commissioning of required capacity at the end of this year.
- The investment component r includes *capital investments required for expansion of power plants only* (electric networks are not considered).
- Auxiliary power consumption, network losses, and taxes are not taken into consideration.
- The installed capacity of EPS N is supposed to increase (following the electricity consumption rise) at a *constant* annual pace λ:

$$N_t = N_0(1 + \lambda)^t. \tag{6.4}$$

- Consideration is given to *the long run* of EPS expansion, which exceeds essentially the service life of power plants T_L and, more so, the credit or investment recovery periods T_R.

[1] The material of the section is based on and, to a great extent, repeats the content of Sect. 3.2 of the monograph [19].

In the Appendix the equations for the investment component r are derived. Below only their final form is presented (the numbers of financing mechanisms are the same as in the previous section).

For the regulated *self-financing* monopoly (Mechanism 1) the electricity tariff to provide the expansion of generation capacities at a pace λ is determined by the expression

$$p_1 = \frac{k}{h}\lambda + c, \tag{6.5}$$

where k is specific capital investments (\$/kW) and h is the annual number of capacity utilization hours of power plants (h/y) at p_1 and c measured in \$/kWh.

The simple form of expression (6.5) is explained by the fact that under self-financing the annual volume of capital investments K_t required to provide annual increase in capacity ΔN_t,

$$K_t = k\Delta N_t = k\lambda N_{t-1},$$

is allocated to the output of power plants available by the end of the previous year $Q_t = hN_{t-1}$,

$$r_1 = \frac{K_t}{Q_t} = \frac{k\lambda N_{t-1}}{hN_{t-1}} = \frac{k}{h}\lambda. \tag{6.6}$$

Expression for the tariff under *Mechanism 2* (for the regulated monopoly that borrows *credits* or for the single-buyer market) appears to be much more complicated:

$$p_2 = \frac{k}{h}\frac{\sigma}{1-(1+\sigma)^{-T_R}}\left[1-(1+\lambda)^{-T_R}\right] + c \tag{6.7}$$

or, if to transform it into the expression

$$p_2 = \frac{k}{h}\frac{\sigma(1+\sigma)^{T_R}}{(1+\sigma)^{T_R}-1}\frac{(1+\lambda)^{T_R}-1}{(1+\lambda)^{T_R}} + c, \tag{6.7a}$$

where T_R is the credit or investment recovery period (years) and σ is the interest on capital (in fractions of unity).

As is seen, in this case the investment component of tariff represents an annual repayment volume of credits or investments (borrowed or invested in the previous years) that is allocated to the annual output of EPS and depends not only on the pace of expansion λ, but also on the recovery period T_R and interest on capital σ.

The second fraction in expressions (6.7) and (6.7a) represents a rather well-known and widely applied *capital recovery factor* (CRF) [23]:

$$\text{CRF} = \frac{\sigma}{1-(1+\sigma)^{-T_R}} = \frac{\sigma(1+\sigma)^{T_R}}{(1+\sigma)^{T_R}-1}. \tag{6.8}$$

It was obtained on the assumption that the borrowed capital is recovered with an interest on capital σ by equal annual parts during T_R years. By multiplying the total borrowed capital by this dimensionless factor we will obtain an annual amount to be paid back. If we multiply the factor by the number of repayment years T_R, we will know how much the total amount to be paid back exceeds the initially borrowed amount because of interest σ. This factor (CRF) will also be used in the equation of electricity price under the competitive market to be considered below.

Expression (6.8) with a recovery period assumed to be equal to service life ($T_R = T_L$) is used in [23] to estimate fixed costs (that do not depend on operating conditions) of power plants. In fact, investors will seek to recover their capital much earlier (inequality (6.2) in the previous section).

The third fraction (or expression in square brackets) in expressions (6.7) and (6.7a) that contains the pace λ characterizes a progressively increasing debt of monopoly company in the amount of borrowed credits.[2] When deriving Eq. (6.7) or (6.7a) the credits in all years were assumed to be borrowed at an identical interest rate σ and for an identical period T_R. Therefore, this fraction contains the recovery period T_R used in the second fraction (the debt of the company was accumulated over the previous period T_R, while it had already been repaid for the years before).

We did not find the expressions of the form (6.7) or (6.7a) in publications, therefore they are likely to be presented in [19] for the first time. On analogy with CRF, two fractions that are used in these expressions after k/h will be called the *capital recovery factor at expanding generation* (CRFEG).

It should be noted that expressions (6.5) and (6.7) for the tariffs of a monopoly company have the sense of tariffs that are established by regulatory bodies (energy commissions). Such tariffs, on the one hand, provide normal expansion and operation of EPS that belongs to the company (the costs are supposed to include normal profit, depreciation charges, taxes, and other operating costs of the company). On the other hand, with such tariffs the company will not have a monopoly profit (at one-type power plants the tariffs will naturally be average throughout an EPS or a company).

Under the single-buyer market the expression (6.7) represents an *average weighted* tariff for electricity bought by the Purchasing Agency from *operating* and *new* producers (PGCs and NPPs). Here it is but again supposed that all contracts for the construction of new power plants assume one and the same investment repayment period T_R and identical interest on capital σ.

For the *competitive market* (Mechanism 3) the price in the electricity wholesale market that provides recovery of private investments and operating costs will have the form

$$p_3 = \frac{k}{h} \frac{\sigma}{1 - (1 + \sigma)^{-T_R}} + c, \tag{6.9}$$

[2] A similar situation will also occur under the single-buyer market; however, it is simpler to explain the essence of the equations by the example of the monopoly borrowing credits.

or after transformation into the expression

$$p_3 = \frac{k}{h} \frac{\sigma(1+\sigma)^{T_R}}{(1+\sigma)^{T_R} - 1} + c, \tag{6.9a}$$

where all notations remain the same. Taking into account the fact that private investments under the competitive market should be recouped at the expense of selling electricity generated by the constructed power plant itself, expressions (6.9) and (6.9a) are written for each power plant individually (and do not contain a pace λ). Here the power plant is assumed to be of the same type as under the regulated markets and with the same technical and economic indices (excluding, probably, costs that can be lower than under the monopoly).

The multiplier at $\frac{k}{h}$ in (6.9) and (6.9a) represents the CRF that has already been considered in (6.8). On the whole, these equations are simpler than for Mechanism 2, and the price calculated by them has the sense of the *minimum* wholesale electricity market *price* at which the investments will be recovered in the period T_R at an interest rate σ. This price should naturally be maintained during the whole recovery period.

The investment component r has a different form in expressions (6.5), (6.7), and (6.9). Therefore, it is appropriate to make their qualitative and quantitative analysis. However, expressions (6.7) and (6.9) are quite complicated, and therefore it is desirable to simplify them and make more illustrative. This can be done by expanding the binomial function $(1+\sigma)^{T_R}$ or $(1+\lambda)^{T_R}$ into series by binomial formula [75] and use only the first, most significant members of the series. Without demonstrating detailed transformations we can say that by using two first members of such series it is possible to obtain

$$(1+\sigma)^{T_R} \approx 1 + \sigma T_R \quad \text{and} \quad (1+\lambda)^{T_R} \approx 1 + \lambda T_R. \tag{6.10}$$

Using these equations and expressions (6.7a) and (6.9a) it is easy to obtain the *approximated formulas for tariffs and prices*. For the regulated monopoly borrowing credits and single-buyer market, we will have

$$p_2 = \frac{k}{h} \lambda \frac{1+\sigma T_R}{1+\lambda T_R} + c, \tag{6.11}$$

and for the competitive market

$$p_3 = \frac{k}{h}\left(\sigma + \frac{1}{T_R}\right) + c. \tag{6.12}$$

These expressions along with expression (6.5) for a self-financing monopoly, that did not require any simplifications, have become much simpler to analyze and compare. For example, expression (6.11) has appeared to be sufficiently close to (6.5): now it has an additional fraction of simple form that contains the values of σ, λ, and T_R.

A detailed *qualitative analysis* of formulas for tariffs and prices is presented in [19]. Here we will indicate its most interesting results.

Analysis and comparison of expressions (6.5) and (6.11) show the following trends and distinctive features as applied to the *regulated monopoly* and single-buyer market:

1. At $T_R = 0$, expression (6.11) gets identical to expression (6.5). This means that self-financing is equal to immediate recovery of credits and has quite a real economic sense (i.e., corresponds to the actual financing mechanism).
2. At $\lambda = \sigma$, the numerator and denominator of the fraction in expression (6.11) are reduced and the expression again becomes identical to expression (6.5). Thus, the conclusion can be made that if credits are borrowed at an interest rate equal to the expansion pace, it makes no difference what method of financing to apply. Here the investment component r will be the same under self-financing and borrowing, and in case of borrowing the credit period will not matter (numerator and denominator that contain T_R in expressions (6.7) and (6.11) are canceled and T_R disappears). This conclusion should be treated carefully. It is true for the EPS expanding for a long time at a constant pace λ. If it is to be imagined that at some time instant EPS ceases to expand, then, in the event that company borrows credits, it will have debts to banks and these debts will have to be paid back during a certain period, by including them in the investment component. Under self-financing there will be no such debts, and electricity tariff during the period of "aftereffect" will be lower than at borrowing credits. This circumstance is reflected neither in formulas (6.5) and (6.11) nor in expressions (6.7) and (6.7a) due to the assumption on constant pace λ, which was made to derive them.
3. If $\lambda < \sigma$, in expression (6.11) the fraction

   $$\frac{1 + \sigma T_R}{1 + \lambda T_R} > 1,$$

   i.e., tariff p_2 at borrowing credits will be higher than tariff p_1 at self-financing. This means that at $\lambda < \sigma$ it is appropriate not to borrow credits but expand generation capacities by self-financing. At the same time this means that at slow paces of expansion as is the case in the USA and Western Europe, it is inappropriate to transform the regulated monopoly into a single-buyer market in which financing mechanism is similar to borrowing credits.

   However, if, vice versa, $\lambda > \sigma$, the indicated fraction will become less than unity and the relationship between the tariffs will be $p_2 < p_1$. Hence, it is more profitable to develop monopoly at the expense of credits or to shift to the single-buyer market. This situation is characteristic of vigorously developing countries: China, India, Brazil, and others.
4. If $\sigma = 0$, in the regulated monopoly the tariff at borrowing credits will always be lower than the tariff at self-financing:

   $$p_2 = \frac{k}{h} \frac{\lambda}{1 + \lambda T_R} + c < p_1,$$

i.e., expansion is always profitable at the expense of interest-free credits and it is better to take them at the longest period possible $T_R = T_L$. Economically, this circumstance is quite obvious.

Now let us compare (also qualitatively) the *tariffs under financing Mechanism 2* p_2 (6.7) *with prices* p_3 that are necessary to recover private investments *under the competitive market* (6.9). As is seen, the investment component under the competitive market in expression (6.9) contains CRF (6.8):

$$p_3 = \frac{k}{h}\mathrm{CRF} + c. \tag{6.9b}$$

The tariff in Mechanism 2 (6.7) also contains CRF, but it is multiplied by an additional factor that includes the pace of expansion λ and reflects the fact that in Mechanism 2 investments are allocated to the output of the entire EPS but not to the output of only one newly constructed power plant as is the case under the competitive market:

$$p_2 = \frac{k}{h}\mathrm{CRF}\left[1 - (1+\lambda)^{-T_R}\right] + c. \tag{6.7b}$$

It is easy to see that the expression in square brackets is always lower than unity at positive λ and T_R (at $\lambda = 0$ it vanishes). This can be achieved, for example, by using the approximated equation (6.10):

$$1 - (1+\lambda)^{-T_R} = 1 - \frac{1}{(1+\lambda)^{T_R}} \approx 1 - \frac{1}{1+\lambda T_R} < 1. \tag{6.13}$$

Thus, the following important conclusion can be made:

5. At *identical* interests on capital σ and credit and private investment repayment periods T_R, the investment components of tariff in the regulated monopoly and the single-buyer market is *always lower* than the similar component of price required to recover investments under the competitive market. This difference, though, as is seen from (6.13), decreases with increasing λ and T_R.
   Meanwhile, as has been mentioned many times, due to the increasing financial risk in the competitive market, the interest on capital σ rises as compared to the credits in the regulated monopoly or investments in the single-buyer market. This is reflected in the value of CRF (6.8). The analysis of expression (6.8) shows that the higher the interest rate σ, the larger the CRF. Hence in the competitive market (6.9b) the CRF will be higher than under Mechanism 2 (6.7b), and another conclusion can be drawn:
6. *Higher financial risk* under the competitive market *additionally increases the electricity price* required to recover private investments in power plants as compared to the tariff under the regulated monopoly or the single-buyer market. This circumstance makes the conclusion drawn in point "5" even more convincing.
   Using the approximated formula (6.12) we can also *compare* the *tariff* p_1 (6.5) in the regulated self-financing monopoly with the price p_3 in the competitive

market. Based on the comparison of (6.5) and (6.12), another conclusion can be made:

7. Tariffs in the regulated self-financing monopoly will be knowingly lower than the prices required to recover investments under the competitive market if the equality

$$\lambda < \sigma + 1/T_R \tag{6.14}$$

is satisfied, where σ corresponds to the higher-risk conditions ($\sigma = 0.12$–0.20). This means that at the recovery period $T_R = 10$–15 years the competitive market can be more efficient than the regulated self-financing monopoly at a pace of the generation capacity expansion $\lambda > 0.18$–0.30, which is virtually unreal.

Let us now make a *quantitative analysis* of tariff and price values for different relationships among λ, σ, and T_R, which can occur in different countries. The analysis will be based on expressions (6.5), (6.7), and (6.9) that are obtained in Appendix 1 at discrete representation of EPS expansion with annual time intervals. Again, EPS will be assumed to consist of one-type power plants with identical technical and economic indices.

Based on the fact that the indicated expressions vary only in multipliers at $\frac{k}{h}$, we will denote these multipliers by A_1, A_2, and A_3, respectively:

$$A_1 = \lambda, \tag{6.15}$$

$$A_2 = \frac{\sigma}{1-(1+\sigma)^{-T_R}}\left[1-(1+\lambda)^{-T_R}\right], \tag{6.16}$$

$$A_3 = \frac{\sigma}{1-(1+\sigma)^{-T_R}}. \tag{6.17}$$

Let us remind that multiplier A_3 represents CRF, therefore expression (6.17) coincides with (6.8). The same CRF enters into expression (6.16) for A_2, which was called capital recovery factor at expanding generation (CRFEG). Table 6.2 presents the CRF values depending on σ and T_R, which are calculated by formula (6.8) which is identical to (6.17). These values of A_3 will be used for further analysis.

Table 6.2 The values of *CRF—capital recovery factor* (multiplier A_3)

T_R	σ						
	0.00	0.03	0.05	0.08	0.10	0.15	0.20
5	0.2000	0.2184	0.2310	0.2505	0.2638	0.2983	0.3344
10	0.1000	0.1172	0.1295	0.1490	0.1627	0.1993	0.2385
15	0.0667	0.0838	0.0963	0.1168	0.1315	0.1710	0.2139
20	0.0500	0.0672	0.0802	0.1019	0.1175	0.1598	0.2054
25	0.0400	0.0574	0.0710	0.0937	0.1102	0.1547	0.2021
30	0.0333	0.0510	0.0651	0.0888	0.1061	0.1523	0.2008

Note that at $\sigma = 0$

$$\text{CRF}_{\sigma=0} = \frac{1}{T_R}. \tag{6.18}$$

This relationship is based on a sufficiently obvious fact: if credit or investments are recovered without interests by equal annual amounts, then the $1/T_R$ part of credit is repaid annually. It can also be obtained by finding the limit of expression (6.8) at $\sigma \to 0$ or if to substitute $\sigma = 0$ into expression (6.12).

Table 6.3 presents the values of multiplier A_2 (CRFEG), which depend not only on σ and T_R, but also on λ. Using the values from Tables 6.2 and 6.3, we will first check the validity of some conclusions made on the basis of qualitative analysis and comparison of expressions (6.5), (6.7), and (6.9).

An analysis of Table 6.3 shows that at all values of T_R the value of A_2 appears to be identical at $\lambda = \sigma$ and equal to these values of λ and σ ($A_2 = \sigma = \lambda = A_1$). This confirms the *thesis* "2": "… if credits are taken at an interest rate σ, equal to pace of expansion λ, it does not matter what method of financing (self-financing or borrowing credits) should be used in the regulated monopoly."

Further, if we proceed from some value of $A_2 = \sigma = \lambda$ (at any T_R) and follow the values of A_2 in the line with the fixed σ, it will be seen that:

- The values of A_2 become larger than λ ($A_2 > \lambda = A_1$) as the value of λ decreases ($\lambda < \sigma$).
- The values of A_2, on the contrary, turn out to be smaller than λ ($A_2 < \lambda = A_1$) as the value of λ increases ($\lambda > \sigma$).

This fact confirms the *thesis* "3" that at $\lambda < \sigma$ the tariff p_1 (6.5) will be lower than p_2 (6.7) and in the regulated monopoly it is expedient to self-finance generation capacity expansion, and at $\lambda > \sigma$, on the contrary, $p_1 > p_2$ and expansion based on credits is reasonable. At the same time, at $\lambda > \sigma$ the single-buyer market that is equivalent in this case to the regulated monopoly with crediting (the same Mechanism 2 of financing new power plants) becomes effective.

Let us verify the important *thesis* "5" affirming that at the same interest on capital σ and periods of credit and private investments repayment T_R, the investment component of tariffs in the regulated monopoly or the single-buyer market (Mechanism 2 of financing) is *always lower* than the price component in the competitive market which is necessary for investment recovery (Mechanism 3). This requires comparison of the numbers of Table 6.2 (A_3) and Table 6.3 (A_2) at the same values of σ and T_R. It can be seen that this is really so, especially at small values of the period of payback T_R and the rate λ. For example, at $\sigma = 0.1$ and $T_R = 10$ years:

- The value of A_3 (and CRF) in Table 6.2 equals 0.1627.
- The value of A_2 (Table 6.3) equals 0.0628 at $\lambda = 0.05$ (2.6 times lower) and 0.1000 at $\lambda = 0.10$ (1.6 times lower).

Difference in the values of A_3 and A_2 decreases with the increasing T_R and λ.

Table 6.3 The values of CRFEG—capital recovery factor at expending generation (multiplier A_2)

T_R	σ	λ					
		0.01	0.03	0.05	0.10	0.15	0.20
5	0.00	0.0097	0.0275	0.0433	0.0758	0.1006	0.1196
	0.03	0.0106	0.0300	0.0473	0.0828	0.1098	0.1306
	0.05	0.0112	0.0317	0.0500	0.0876	0.1161	0.1382
	0.08	0.0122	0.0344	0.0542	0.0949	0.1259	0.1498
	0.10	0.0128	0.0362	0.0571	0.1000	0.1326	0.1578
	0.15	0.0145	0.0410	0.0646	0.1131	0.1500	0.1784
	0.20	0.0162	0.0459	0.0724	0.1268	0.1681	0.2000
10	0.00	0.0095	0.0256	0.0386	0.0614	0.0753	0.0838
	0.03	0.0111	0.0300	0.0453	0.0720	0.0883	0.0983
	0.05	0.0123	0.0331	0.0500	0.0796	0.0975	0.1086
	0.08	0.0141	0.0381	0.0575	0.0916	0.1122	0.1250
	0.10	0.0154	0.0416	0.0628	0.1000	0.1225	0.1365
	0.15	0.0189	0.0510	0.0769	0.1224	0.1500	0.1671
	0.20	0.0226	0.0610	0.0921	0.1466	0.1796	0.2000
15	0.00	0.0092	0.0239	0.0346	0.0507	0.0585	0.0623
	0.03	0.0116	0.0300	0.0435	0.0637	0.0735	0.0783
	0.05	0.0134	0.0345	0.0500	0.0733	0.0845	0.0901
	0.08	0.0162	0.0418	0.0606	0.0889	0.1025	0.1092
	0.10	0.0182	0.0471	0.0682	0.1000	0.1153	0.1229
	0.15	0.0237	0.0612	0.0888	0.1301	0.1500	0.1599
	0.20	0.0297	0.0766	0.1110	0.1627	0.1876	0.2000
20	0.00	0.0090	0.0223	0.0312	0.0426	0.0469	0.0487
	0.03	0.0121	0.0300	0.0419	0.0572	0.0631	0.0655
	0.05	0.0145	0.0358	0.0500	0.0683	0.0753	0.0781
	0.08	0.0184	0.0455	0.0635	0.0867	0.0956	0.0992
	0.10	0.0212	0.0524	0.0732	0.1000	0.1103	0.1144
	0.15	0.0288	0.0713	0.0995	0.1360	0.1500	0.1556
	0.20	0.0371	0.0917	0.1280	0.1748	0.1928	0.2000
25	0.00	0.0088	0.0209	0.0282	0.0363	0.0388	0.0396
	0.03	0.0126	0.0300	0.0405	0.0521	0.0557	0.0568
	0.05	0.0156	0.0371	0.0500	0.0644	0.0688	0.0702
	0.08	0.0206	0.0489	0.0660	0.0850	0.0908	0.0927
	0.10	0.0243	0.0576	0.0776	0.1000	0.1068	0.1090
	0.15	0.0341	0.0808	0.1090	0.1404	0.1500	0.1531
	0.20	0.0445	0.1056	0.1424	0.1835	0.1960	0.2000
30	0.00	0.0086	0.0196	0.0256	0.0314	0.0328	0.0332
	0.03	0.0132	0.0300	0.0392	0.0481	0.0502	0.0508
	0.05	0.0168	0.0383	0.0500	0.0613	0.0641	0.0648
	0.08	0.0229	0.0522	0.0683	0.0837	0.0875	0.0885
	0.10	0.0274	0.0624	0.0815	0.1000	0.1045	0.1056
	0.15	0.0393	0.0896	0.1171	0.1436	0.1500	0.1517
	0.20	0.0518	0.1181	0.1544	0.1893	0.1978	0.2000

Now consider *concrete ratios of multipliers* A_1, A_2, and A_3 for some countries and regions of the world. For this purpose we choose industrially developed countries (Western Europe states, the USA), where the pace of power industry development is relatively slow (in the recent years and the coming future), Russia, and

China. In the first group the paces of power industry development may be assumed to be equal to 1–3% ($\lambda = 0.01$–0.03), in Russia about 5%, and in China the pace is observed to be up to 10–15%.

As to the interest σ and the payback periods T_R for the *credits* in the regulated monopolies or for the investments in the single-buyer market (Mechanism 2 of financing), they undoubtedly differ by country, concrete project, and time of their receipt. However, they are assumed to be equal for all the considered countries. Note that in the regulated markets there is practically no financial risk and credits or investments can be received at a relatively low interest and for a long term. Therefore, the values $\sigma = 0.08$ and $T_R = 20$ years can be taken as representatives to determine the multiplier A_2.

For the *competitive market* (Mechanism 3) it seems more difficult to evaluate the interest σ and the payback period T_R, since the information about terms of private investments is usually considered as confidential. It is not laid open to public; it is difficult to receive, the more so, to generalize it somehow. Despite this fact it is logic to assume that because of a higher risk the conditions of investing in the competitive market will considerably differ from conditions of crediting and investing in the regulated markets (it was discussed above). An investor will decide to invest only if he receives a higher interest σ. Therefore, the values $\sigma = 0.15$ and $T_R = 15$ years may be taken as representative to determine the multiplier A_3.

The multipliers A_1, A_2, and A_3 for the considered countries are determined in Table 6.4 based on the assumed values of λ, σ, and T_R by using Tables 6.2 and 6.3. They characterize a relative value of the investment component of tariffs and prices that is needed to maintain expansion of single-type generation capacities of EPSs at different mechanisms of financing. These multipliers contained in expressions (6.5), (6.7), and (6.9) for electricity tariffs or prices can be compared with each other directly.

Analysis of Table 6.4 allows the following conclusions to be drawn:

1. In the countries with a slow pace of power industry development (Western Europe, USA, Russia) self-financing in the regulated monopolies (the multiplier A_1) decreases the investment component of tariffs in comparison with crediting (the multiplier A_2) by 1.3–1.8 times. As for China with its rapid pace, on the contrary, crediting or transfer to the single-buyer market is more preferable.
2. The investment component of electricity price in the competitive market (the multiplier A_3) is much higher than that of tariffs in the regulated markets in all considered cases. The difference is particularly large in the industrially developed

Table 6.4 The ratio of the investment component in tariffs and prices for different countries and regions of the world

Country, region	$\lambda = A_1$	A_2	A_3
Western Europe, USA	0.01	0.0184	0.1710
	0.03	0.0455	0.1710
Russia	0.05	0.0635	0.1710
China	0.10	0.0867	0.1710
	0.15	0.0956	0.1710

countries, where the increase is 3- to 15-fold in comparison with self-financing. In China, A_3 is 1.7–2.0 times higher than A_2. It means that financing the generation capacity expansion in the regulated electricity market (Mechanism 1 or 2) is always *more effective* than in the deregulated market (Mechanism 3).

The multipliers A_1, A_2, and A_3 show ratios of the investment component in prices and tariffs, therefore it is of concern to estimate its values for different types of power plants (in ¢/kWh) quantitatively. If we denote the investment component by letter r, as in (6.3), its quantitative values in accordance with expressions (6.5), (6.7), (6.9), and (6.15)–(6.17) can be determined by the formula

$$r_i = \frac{k}{h} A_i, \quad i = 1, 2, 3, \tag{6.19}$$

where the notation is the same as before.

Table 6.5 presents these values for the main types of power plants in the European section of Russia's UPS (EUPS). They are calculated based on the following assumptions:

- The multipliers A_1, A_2, and A_3 are taken from Table 6.4 in respect to Russia's conditions at the pace $\lambda = 0.05$.
- The specific investments k are accepted in accordance with Table 5.1.
- The number of installed capacity utilization hours h is taken from Table 5.4 (at $Q_{\text{VIC}} = 900\text{TWh}$) with correction for gas-fired condensing power plants (CPPs) with combined-cycle installations that are more effective than the same CPPs with steam turbine installations presented in Table 5.4.

It can be seen that hydropower plants (HPPs) having high capital investments and small number of utilization hours have a very large investment component. For NPPs and CPPs on coal it is approximately the same. It is much smaller for CPPs on gas with combined-cycle installations that will chiefly be constructed (instead of CPPs on gas with steam turbine installations) in the future. The values obtained for CGPPs should be considered as conditional, because half of the capital investments are supposedly attributed to thermal energy production.

As might be expected, the competitive market (column "r_3") had the highest investment component. At the assumed values of λ, T_R, and σ it exceeds that for Mechanisms 1 and 2 of financing:

Table 6.5 Estimates of the investment component of tariffs and prices for Russia (EUPS, 2010)

Power plant	k (\$/kW)	h (h/year)	r_1 (¢/kWh)	r_2 (¢/kWh)	r_3 (¢/kWh)
HPP	2,200	3,000	3.67	4.66	12.54
NPP	1,650	7,000	1.18	1.50	4.03
CPP (coal)	1,200	5,000	1.20	1.52	4.10
CPP with CCI (gas)	800	5,500	0.73	0.92	2.49
CGPP (gas)	600[a]	5,000	0.60	0.76	2.05

[a] Half of the capital investments in cogeneration power plants (CGPPs) are attributed to thermal energy production.

- For CPPs on gas with combined-cycle installation (CCIs)—by more than 1.5 ¢/kWh
- For NPPs and CPPs on coal—by 2.5–2.8 ¢/kWh
- For HPPs—by 8–9 ¢/kWh

Solely the numbers in column "r_3" of Table 6.5—for Mechanism 3 of financing, when each power plant must cover its own expenditures—may be used directly. In this case they show an *excess of prices in the competitive* wholesale market over operating costs of the corresponding type of power plants, at which private investments will be paid back (at the accepted period T_R with the expected interest σ). In fact, it will be an excess of the wholesale prices that is necessary for *new* power plants over the costs of similar *operating* power plants. Such an excess, as was noted earlier, creates a *price* barrier for entrance of new producers into the competitive electricity market.

In the regulated monopolies and for the single-buyer market the investment component of tariffs will depend on the type and ratios of new (being constructed) power plants. Thereby the values in columns "r_1" and "r_2" will be "weighted" in a certain way. This "average weighted" investment component will depend on the optimal mix (and capacities) of new power plants. As a rule, it will be lower than the value of r_3 even for the most effective combined-cycle installations.

The full value of electricity tariffs or prices p (6.3) depends, in addition, on production costs c, which differ for diverse types of power plants. In the regulated markets (Mechanisms 1 and 2) the average weighted costs of generation will be dictated by the structure of *operating* power plants.

Box 15 Mathematical Expressions for the Investment Component of Electricity Tariffs and Prices, Their Qualitative and Quantitative Analysis

1. The formulas were obtained to calculate investment components under different financing mechanisms for construction of new power plants. These formulas reflect major regularities in their formation:

   (a) Under Mechanism 1 (self-financing with a market organized according to Model 1) the investment component r depends on the generation capacity expansion pace λ.

   (b) Under Mechanism 2 (bank credits according to Model 1 and private investments according to Model 2) r depends on the pace λ, the interest rate σ, and the payback period T_R.

   (c) Under Mechanism 3 (private investments according to Models 3 and 4) r is determined by the values of σ and T_R.

2. Qualitative analysis of these formulas revealed some trends and regularities:

At $\lambda < \sigma$ it is profitable for regulated monopolies to have self-financing, and at $\lambda > \sigma$, vice versa, bank credits or transition to the single-buyer market. At equal σ and T_R the investment component of tariffs under Mechanism 2 is *always lower* than the component of prices required for investment recovery under Mechanism 3. This is explained by the fact that in the regulated markets the investments are allocated to the entire output of *all* power plants in EPS and in the competitive market the investments in a new power plant should be recovered at the expense of generation of this *one* new *power plant only*. Higher financial risk in the competitive market aggravates this trend.

3. Quantitative calculations confirm the general trends revealed in the qualitative analysis:
   a. In countries and regions with a development pace $\lambda = 0.01$–0.05 (Western Europe, the USA, and Russia) self-financing in the regulated monopolies decreases the investment component of tariffs by 1.3–1.8 times as compared to the bank credits. At a very rapid pace $\lambda = 0.10$–0.15 (China), vice versa, the bank credits or the single-buyer market are more preferable.
   b. For Russia the investment component under Mechanism 3 (competitive market) will be higher by 1.5–3 ¢/kWh than under Mechanisms 1 and 2 (markets with regulated electricity prices), depending on the mix of new power plants to be commissioned. The low figure relates to gas-fired CPP with combined-cycle installations.
4. The qualitative and quantitative analyses have shown that financing the new power plants in the competitive market will *always* require *larger price rise* than in the regulated markets according to Models 1 and 2.

6.3 Generation Costs in the Long Run

EPS and its generation capacities are expanding, as is indicated in Sect. 3.1, *in the long run* (in terms of microeconomics), where the firm's production capacities *change* (are not fixed). Duration of the long run can be different depending on specific features of the industry. Correspondingly, it is necessary to consider *long-run* costs and the firms' supply curves, *long-run* curves of purchasers' demand, and *price formation in the long run.* In the competitive (unregulated) market the equilibrium prices that are formed in the long run should specify an *actual* commodity *price* in the developing industry.

As concerns the power industry, as was mentioned earlier, in accordance with the model of market organization the firm may be represented by a regulated monopoly company (VIC), a PGC consisting of several power plants, and even an individual power plant (independent or new power producer) and also different network, distribution, and sales companies not studied here. Let us analyze long-run costs of individual power plants, generation spheres of VICs, and PGCs. Section 6.4 is devoted to price formation in the long run.

6.3.1 Long-Run Costs of Power Plants

For *individual* power plants, as may be supposed, the notion of long-run costs in general has no sense. It is possible to speak only about short-run costs of *operating* and *new* power plants.

The operating (not upgradable) power plants have fixed installed capacities and fixed costs that are independent of the annual electricity production. Hence, they may have only short-run costs addressed in the previous chapter.

Every new power plant is also an independent object designed for a certain installed capacity. If built, it will operate with this invariable capacity and costs that, in terms of microeconomics, should be referred to the short-run ones. However, in comparison with the operating power plants the costs of new ones include an *investment component* in addition to purely operating costs. And this fact will influence participation of a new power plant in the electricity market.

If the power plant operates in the regulated VIC, its operating costs (fixed and variable) will be included in the costs of the VIC's generation sphere and its investments—in the investment component of tariffs for consumers. When the new power plant participates in the single–buyer market, it will supply electricity to the Purchasing Agency at a higher (in comparison with operating power plants) price that takes into account investment recovery at the agreed time T_R with some interest σ. Here the operating costs of and investments in a power plant will be included in the average weighted tariff for consumers with the investment component.

We will face quite a different situation when an individual *new* power plant participates in the competitive electricity market. It will offer the prices that are set based on its full costs including both the operating costs and the investment component. This corresponds to financing Mechanism 3 of power plant construction. The electricity price it can offer (at which its investments will be recouped) will be determined by expression (6.9). The price will naturally be much higher than the costs of similar operating power plants.

The short-run costs of different *operating* and *new* power plants are calculated for the European section of Russia's UPS at 2010 level in Table 6.6 as an example. These costs are used as a basis for establishing prices the plants will offer in the *competitive* market. The costs of new power plants should be considered as *long-run costs* of individual power plants of the corresponding type.

The number of utilization hours h and the investment component r_3 (for Mechanism 3 of financing) are assumed the same as in Table 6.5. The short-run average

Table 6.6 Short- and long-run costs of individual power plants (EUPS of Russia, 2010), ¢/kWh

Power plant	h (h/year)	Short run	Long run	
		SATC	r_3	*LAC*
HPP	3,000	1.17	12.54	13.71
NPP	7,000	2.28	4.03	6.31
CPP (coal)	5,000	3.32	4.10	7.42
CPP (gas) with CCI	5,500	2.44	2.49	4.93
CGPP (gas)	5,000	2.81	2.05	4.86

total costs (*SATC*) are taken from Table 5.9 with correction of their values for CPPs on gas with CCIs (in Table 5.9 these CPPs were assumed to have steam turbine installations). The long-run average costs (*LAC*) are determined as a sum of the short-run costs and the investment component:

$$LAC = SATC + r_3. \tag{6.20}$$

The long-run costs of power plants (or short-run costs of new power plants) are seen to be much higher than the short-run costs, particularly for capital-intensive HPPs. They increase to the least extent for CPPs on gas with CCIs. As before, the long-run costs of CGPPs should be considered conditional, because half their capital investments are attributed to thermal energy production.

Thus, the short-run costs of *new* power plants with the investment component should be considered as long-run costs of individual power plants in the *competitive* electricity market. The process of participation of new power plants in the competitive wholesale market will be illustrated in Sect. 6.4.

6.3.2 Long-Run Costs of VIC's Generation Sphere

Expansion of generation capacities in the *regulated* VIC (Model 1) is financed by including investments in the investment component of consumer tariffs. As is shown in Sects. 6.1 and 6.2, two ways and mechanisms of financing new power plants are possible—self-financing (Mechanism 1) and construction by using bank credits (Mechanism 2). In both cases the VIC costs increase by this investment component.

Hence, in the long run the *investment component* is added to the *short-run costs* of VIC. If account is taken of generation sphere only without investments in the electric networks of EPS, the key features of this component under Mechanisms 1 and 2 will be such as was discussed in the previous section.

In general, the *long-run costs* of VIC can be expressed as

$$LAC_{\mathrm{VIC}} = SATC_{\mathrm{VIC}} + r, \tag{6.21}$$

where $SATC_{\mathrm{VIC}}$ is the short-run average total costs of VIC and r is an investment component. In the long-run costs there is no fixed component (all costs are variable), therefore it is written only that these are average (per 1 kWh) costs (*LAC*).

The sense of the long-run costs of VIC *complies rather well* with their notion in the theory of microeconomics. They contain investments required for company development that is naturally optimized. Therefore, in the course of introduction of more and more advanced technologies for electricity generation they will decrease with the increasing size of EPSs. Or if the generation technologies remain unchanged, the specific costs in the sphere of electricity transmission and distribution will decrease under the influence of technological progress. A similar positive effect can be achieved at the interconnection of EPSs of the considered VIC with neighboring EPSs (see Sect. 2.2). If we neglect such external factors as inflation, fuel price rise, etc., the *long-run costs of VIC can be expected to decrease* with its capacity (annual production) growth, at least, if expression (6.21) is referred to VIC as a whole, but not only to the generation sphere.

Decrease in the long-run costs of VIC with its growing production means that it is a *natural monopoly* and its long-run *marginal costs* (*LMC*) *are lower than the average ones* (*LAC*). This fact is essential at the state regulation of natural monopolies.

In the case of regulation the tariffs for VIC should be established on the basis of *long-run* costs because of the necessity to take into consideration the company's expenses for EPS expansion. Then the question arises, what costs—average or marginal—should be applied. Theoretically, the market equilibrium is believed to be optimal at the intersection of the producers' supply curve that represents *marginal* costs with the purchasers' demand curve. However, if the tariff for the natural monopoly is established based on the long-run *marginal* costs, its *average* costs will not be fully compensated and the company will go bankrupt. Therefore, the regulated monopoly will be an "exception from the rule," i.e., tariffs there are established on the basis of long-run *average* costs (*LAC*).

It should be noted that the same refers to the short-run costs of power plants. The short-run marginal costs (*SMC*) of power plants are lower than their average total costs (*SATC*). Therefore, under regulation their tariffs should be fixed (if EPS does not expand) at the level of *SATC* (rather than *SMC*). And in the competitive wholesale market the power plants must also participate with these total, rather than marginal, costs.

It is difficult enough to graphically represent the curve of long-run costs of VIC as in Fig. 3.4. The figure was purely illustrative and showed only the basic meaning of long-run costs of some ("typical") firm. Meanwhile, the power-generating VIC has certain essential features.

First, VIC (and EPS) is continuously (annually) expanding, and in a facility-by-facility way. It is impossible to clearly distinguish some options of its expansion that correspond to increasing output of VIC Q_{VIC}. Such options are numerous, they are searched in the process of EPS expansion optimization for the future time horizon, and the best one is chosen from them for each future period (year).

Second, the curves of total short-run costs of VIC, as is shown in Sect. 5.4, have the form that differs from the U-shaped form that is supposed for "typical" firms, and Fig. 3.4 is constructed based on it. The curves for $SATC_{VIC}$ (see Fig. 5.7) have a descending shape and reach a minimum at the maximum annual electricity generation Q_{max}. (Such a shape is apparently an additional confirmation that VIC has properties of the natural monopoly.) The author has not managed to graphically illustrate the process of adding the investment component r to $SATC_{VIC}$ and extrapolation to higher production capacity of VIC.

In this context we will not try to graphically interpret long-run costs of VIC and restrict to their formal presentation as expression (6.21). Such values of LAC_{VIC} should be applied to set electricity tariffs for VIC by the regulatory body (expenses on electricity transmission, distribution and sale, and also normal profit being added).

6.3.3 Long-Run Costs of PGCs

The sense and use of long-run costs of PGCs greatly depend on the model of electricity market organization.

In the single-buyer market PGCs supply electricity to the Purchasing Agency by long-term contracts with regulated prices (tariffs). New power plants constructed by some PGCs supply electricity at higher (in comparison with operating power plants) prices that ensure investment recovery (at the agreed period T_R and the interest σ). For PGC as a whole the costs of operating and new power plants are *averaged* by analogy with the costs of VIC's generation, and the revenues gained owing to electricity sale to the Purchasing Agency completely cover them. In this case the *weighted average* costs of operating and new power plants may be taken as the *long-run* costs of PGC. In addition to operating costs they will include an investment component that is represented by the sums of the annual return of investments in new power plants. Formally for the single-buyer market the long-run costs of PGC may be written as expression (6.21), substituting the index "VIC" by "PGC" in it.

Actually in the single-buyer market the long-run costs of PGC are not so important, since this market is not "classical," in which the prices are fixed at the intersection of the curves of producers' supply and purchasers' demand. In the single-buyer market PGCs compete with one another for concluding contracts with the Purchasing Agency for electricity supply from *operating* power plants and separately participate in tenders for construction of *new* power plants. The processes of current operation and expansion of the company are separated in this case. Therefore we will not go into greater detail in the concrete type and quantitative estimates of the long-run costs of PGCs as applied to the considered market model.

In the *competitive* electricity market the situation concerning the long-run costs of PGCs is much more intricate. Here the EPS properties considered in Sect. 2.3 start to display:

- High capital intensity and long periods of power plant construction
- Facility-by-facility EPS expansion

Capacity of PGC increases by constructing a new power plant (if we abstract from possible cases of updating the operating power plants). Each new power plant is an individual (independent) *investment project* that requires substantiation of its effectiveness and financing. As is discussed in Sect. 6.1, the financing mechanism under the competitive market radically differs from such mechanisms in the regulated monopolies and the single-buyer market. The generating company, having the possibility to invest its available capital in *any* branches of the economy, will *individually* and equally assess financial efficiency of the new power plant project as an alternative for investing. A new power plant will be effective for it, only if the investments are paid back in the acceptable period T_R with the desired interest σ owing to sale of electricity generated by *this* power plant (this case was already discussed in Sect. 6.1).

Hence, in the competitive market the sphere of power plant construction by PGCs turns out to be *separated* from the sphere of electricity production by operating power plants. The short-run costs of the latter have nothing to do with investments and costs of a new power plant, and the notion of long-run costs for PGC as a whole, therefore, makes no sense. The *short-run costs of new power plants* that include an investment component need to be considered instead. The situation proves to be identical to NPPs constructing one new power plant.

If the short-run costs of a new power plant built by PGC are interpreted as the long-run costs of this company, expression (6.20) will be true for the latter. The concrete value of LAC_{PGC} will depend on the type of the new power plant, which was illustrated in Table 6.6.

The PGC will offer in the competitive wholesale electricity market the price that is set based on such long-run costs, or rather the short-run costs of new power plants being constructed by the company.

Box 16 Long-Run Costs of Individual Power Plants, VICs, and PGCs

1. *For individual power plants* the notion of long-run costs makes no sense. Instead, it is necessary to consider *short-run costs of new* power plants that include an investment component. The latter for the market organized according to Models 1 and 2 will be in the investment component of tariffs for consumers. In the *competitive* market (Models 3 and 4) the owner of a new power plant (for example, NPP) will offer the electricity price in the long-run market that is set on the basis of costs including investments in this power plant.
2. The sense of long-run costs for *developing* VICs complies rather well with their notion in the theory of microeconomics. The long-run costs of VIC along with the electricity production costs include an investment component. Their average value LAC_{VIC} is determined by dividing the long-run

costs by the total electricity output of the company (from all operating and new power plants).
In view of the "economies of scale," which are typical of EPS, the LAC_{VIC} dependence of the electricity production volume will have a descending form. Long-run marginal costs (LMC_{VIC}) will be lower than average ones. Therefore, electricity tariffs for VIC should be established by the regulatory body at the level of the long-run *average* costs LAC_{VIC} (not LMC_{VIC}), for the company not to be a loser.

3. For PGCs the sense and use of long-run costs depend on the electricity market model:

   a. For the single-buyer market the sense of long-run costs of PGC is almost the same as for VIC, i.e., they contain weighted average electricity production costs and an investment component. This market, however, is not "classical" and the long-run costs of PGC are not involved in the wholesale price formation.
   b. In the *competitive* market, due to a particular financing mechanism for construction of new power plants (Sects. 6.1 and 6.2), PGC will take part in the long-run market *separately* with its operating and new power plants. For PGC to ensure its expansion, the price to be offered by the company in the competitive market should be based on *short-run costs* of new power plants rather than long-run costs of PGC as a whole. The situation turns out to be similar to that described earlier for NPP that constructs one new power plant. In the competitive market the short-run costs of new power plants constructed by PGC should be considered as its long-run costs (or instead of them).

4. The values of short-run costs of *new* power plants of various types, which can be interpreted as *long-run* costs of individual power plants or PGC in the competitive market, are illustrated by the example of the European section of Russia's UPS for 2010.

6.4 Price Barrier for New Power Plants in the Competitive Market

The analysis of short-run costs of power plants and generating companies that was carried out in Chap. 5 and the long-run costs in Sect. 6.3 makes it possible to consider price formation in the *competitive wholesale* market in the long run, i.e., in the process of EPS expansion. To be more precise and convincing, it will be shown again on the example of the European section of Russia's UPS for 2010 level, when it is planned to terminate electricity price regulation.

6.4.1 Initial Principles, Conditions, and Assumptions

It seems necessary to preliminarily present some explanations:

1. In the long-run both *operating* power plants and PGCs and *new* power plants constructed by NPPs or by existing PGCs will participate in the competitive wholesale market. The financing mechanism for new power plants in the competitive market (Mechanism 3) will be *the same* for NPP constructing one (its first) power plant and for existing PGCs. In any case the financial efficiency of a new power plant should be assessed *individually* on the assumption that the investments will be recovered at the expense of its electricity production.
   Therefore, in the *competitive market* the *short-run costs* of *new* power plants that will be constructed (or are planned for construction) by PGCs should be considered as their *long-run costs*. These costs involve an investment component by analogy with new power plants constructed by NPPs. Further, when analyzing the wholesale market prices in the long run, consideration will be given to costs of new power plants of different types assuming that they can be constructed by the existing PGCs or NPPs, i.e., *by any investors*.
2. In Russia due to the "gratuitous" *privatization* of the power industry in the early 1990s the costs of *operating* power plants *do not include any repayments* for the earlier investments. Therefore, the existing power plants will offer in the competitive wholesale market the prices that are fixed on the basis of their short-run costs. The latter are naturally much lower than the costs of new power plants.
3. The electricity consumption growth in the long run will be simulated by the *shift* of the *demand curve of consumers to the right* (to the region of higher electricity production). The slope of this curve depending on demand elasticity will be chosen in the expert way and assumed invariable.
4. As in Sect. 5.3, it will be assumed that *all the operating* power plants of the *same type* are combined to form corresponding PGCs that are treated jointly. The technical and economic indices of various power plants, their total installed capacity, and annual electricity generation at 2010 level are taken the same as in Chap. 5. This makes it possible to apply the results of calculations described earlier.

The other explanations will be given in the course of material presentation.

6.4.2 Comparison of Costs of Operating and New Power Plants

Figure 6.2 presents short-run costs of operating and new power plants in the European section of Russia's UPS at 2010 level. CGPPs on coal are not studied because of their small share—only 2% (see Table 5.1). The costs of only operating power plants are shown for gas-fired CPPs with steam turbines, since it is planned to construct all new CPPs on gas with combined-cycle installations.

The costs of *operating* power plants are taken in accordance with Table 5.9 (ATC_i) and Table 6.6 ($SATC$). In so doing the costs of CPPs on gas with steam turbine

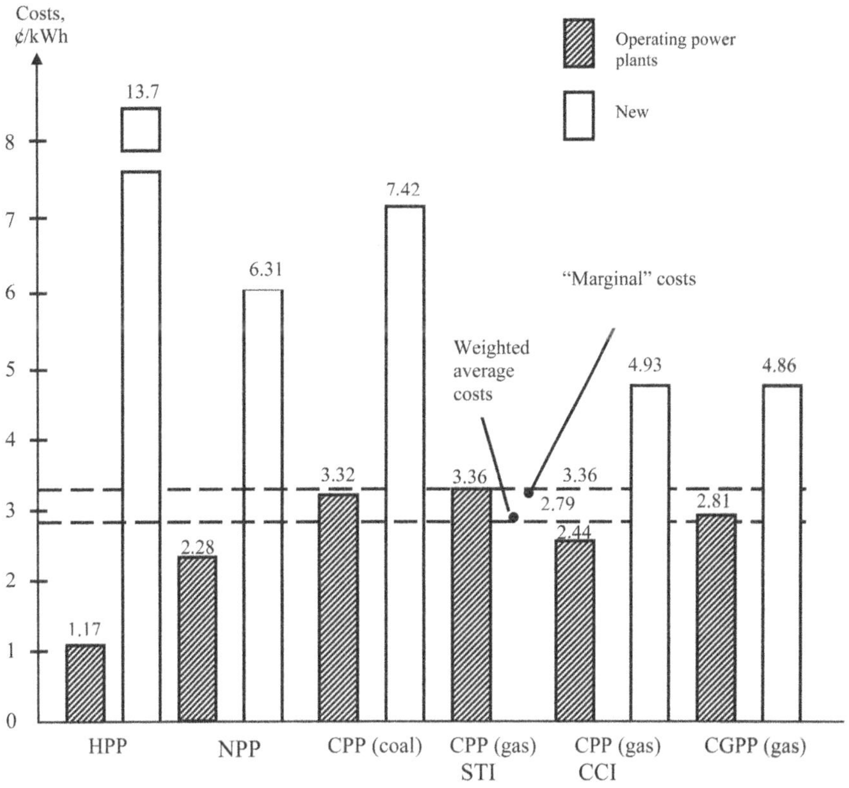

Fig. 6.2 Comparison of costs of operating and new power plants (EUPS of Russia, 2010, 900 TWh)

installations (STIs) are borrowed from Table 5.9 and those with CCIs from Table 6.6. The costs of *new* power plants fully correspond to *LAC* in Table 6.6. Note that the investment component r_3 in the costs of new power plants was determined for the period of investment payback $T_R = 15$ years and the interest on capital $\sigma = 0.15$.

The weighted *average total* costs for the whole EUPS (2.79 ¢/kWh) at an electricity consumption of 900 TWh/year that are assumed in accordance with Table 5.8 (ATC_{VIC}) and Fig. 5.7 and the "marginal" costs (3.36 ¢/kWh), which will underlie price formation in the competitive wholesale market, are shown by dashed lines in Fig. 6.2. The term "marginal" means the *highest* costs of the least effective power plant type. In this case it will be costs of operating CPPs on gas with STIs.

As was indicated in Sect. 5.4, in the *short run* under *regulation of prices* (Models 1 and 2) the wholesale prices (tariffs) will be fixed at the level of average total costs (2.79 ¢/kWh) and *without their regulation* (Models 3 and 4) the equilibrium prices in the wholesale market will be established at the level of "marginal" costs (3.36 ¢/kWh). This price rise will form a producers' surplus in all the remaining types of power plants. This situation will continue until consumer demand increases and construction of new power plants becomes necessary.

The costs of new power plants, as is seen from Fig. 6.2 (and was shown earlier in Table 6.6), are much higher than those of similar operating power plants. In fact, they should be identified with *necessary prices* of the wholesale market, at which the investments will be paid back at the time period T_R with the interest σ, assumed in calculation of the investment component r_3. In other words, in order to attract private investments in construction of new power plants of some type the prices in the *competitive* wholesale market should exceed such costs in the *long run*. This fact should naturally lead to *even greater* price rise of the wholesale market *in the long run* as against the marginal prices in the short-run market. Otherwise, new power plants will not be built.

6.4.3 Price Barrier in the Long Run

Consider now the wholesale electricity price formation in the long run, when construction of new power plants is required. We will illustrate price formation by the same example of the European section of Russia's UPS. According to the resolutions of the Russian Government, the electricity prices will not be regulated completely at the end of 2010. In this context the situation in the competitive wholesale market in EUPS in 2010 that was described in Sect. 5.4 (Fig. 5.8) is assumed to be starting and price formation will be analyzed for the next period.

Figure 6.3 presents a step curve of the average total costs (*ATC*) of PGCs which is taken from Fig. 5.8. The curve should be treated as a supply curve of producers S at electricity consumption in EUPS $Q_{EUPS} = 900$ TWh/year. As was shown in Chap. 5, producers in the power industry (power plants, PGCs) must offer in the competitive market the price that is fixed based on their *total* costs (rather than marginal, as is the case in other industries). The figure was constructed on the assumption that 900 TWh is the *maximum possible* annual production of operating power plants (though basically it can be higher). Therefore, at the given value of Q the supply curve S transfers to the vertical range.

By analogy with Fig. 6.2, the weighted average total costs for the whole EUPS (2.79 ¢/kWh) and the "marginal" costs (3.36 ¢/kWh) are shown by dashed lines in Fig. 6.3. Besides these, the figure presents costs of *new* power plants: CPPs on gas with CCIs (4.93 ¢/kWh), NPPs (6.31 ¢/kWh), and CPPs on coal (7.42 ¢/kWh). The costs of new HPPs and CGPPs are not shown to simplify the figure.

The straight line D_1 presents *consumers' demands* in 2010 before termination of price regulation (assume that until that time they were fully regulated, though the prices were deregulated gradually starting in 2007). Point A corresponds to *solvent* demand at the tariff 2.79 ¢/kWh. The tariffs for different types of power plants (PGCs) were differentiated in accordance with their costs and included only normal profit paid to shareholders.

After the price deregulation, the demand/supply equilibrium will be achieved at point B. The electricity price in this case will soar to the marginal one, i.e., to the

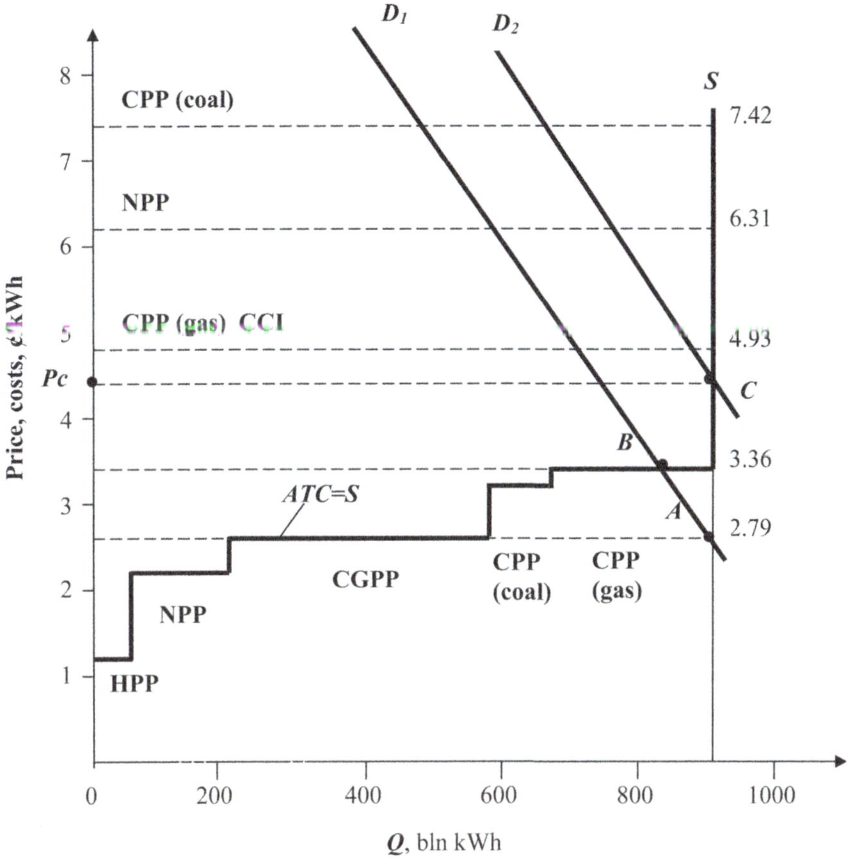

Fig. 6.3 Price formation in the long run in the wholesale market of EUPS (after 2010)

level of costs of PGC with CPPs on gas. Demand will correspondingly fall. At the electricity price corresponding to point B (3.36 ¢/kWh):

- The remaining consumers will bear higher expenses to purchase electricity.
- PGCs with CPPs on gas will receive normal profit, the same as at regulation, i.e., they gain nothing.
- PGCs with the other types of power plants will start receiving extra profit (producers' surplus) equal to the difference between price at point B and their average total costs (*ATC*).

Hence, without price regulation additional expenses of consumers will be used, as was already indicated in Sect. 5.4, for payment of the extra profits to PGCs with more effective power plants. It is unlikely that such a situation can be thought normal.

Consider now the *situation with increasing demand of consumers* for electricity and the arising necessity for EPS expansion (in our case EUPS). At the higher demand D_2 the equilibrium will be attained at point C on the vertical range of the supply curve S. The electricity price will rise to the level p_c. Consumer demand will be satisfied in the amount of maximum possible (as was assumed) annual production 900 TWh. This situation is characterized by:

1. Even higher (than at D_1) expenses of consumers for electricity purchase.
2. Receipt of superprofits by PGCs with CPPs on gas. It will be a monopoly profit that equals the difference between the price p_c and their total costs.
3. Increase in the superprofits of the remaining PGCs. The indicated monopoly profit is added to their producers' surplus.
4. Insufficiency of the price p_c for attraction investments in new power plants—there is a *price* barrier for construction of new power plants and they will not be built.

The described situation is indicative of, first, the *electricity deficit in the market.* The notion "deficit" in this case differs from the traditionally applied one. Sometimes it is said that in the free market deficit is impossible—the prices simply rise, resulting in demand fall and new equilibrium. In fact, some surplus of production capacities is necessary for normal (effective) operation of any market. Only in this case the equilibrium prices will be fixed at the level of marginal costs of producers (at the level of average total costs of the least effective producer in the power industry). If the demand reaches a vertical (inelastic) range of the supply curve of producers, it indicates shortage of production capacities and excess of demand over supply. The electricity price will rise above the costs of the least effective producers with formation of the monopoly profit for them.

The market state, when there is capacity shortage and equilibrium is established on the *vertical* range of the supply curve S, and the electricity price *exceeds* the costs of all producers with receipt of the *monopoly* profit, will be understood as a *deficit*. The term "deficit" will be applied later in this meaning (except the specified cases).

Second, the situation at point C in Fig. 6.3 illustrates a *price barrier* for new producers which occurs in the *competitive* market in the power industry, i.e., the equilibrium price in the wholesale market is insufficient to attract investments in new power plants. In terms of this price barrier the deficit will continue for an indefinite time, say, permanently. If the demand increases still further (the line D shifts to the right) and the price rises to the level of costs for new power plants (and they start to be constructed), the deficit will nevertheless remain. The capacity of any new power plant will make up 2–5% of the total capacity of EUPS, i.e., it will have a minor effect on the demand/supply ratio. And the most important thing is that for construction to be continued (and EPS to be expanded) the price should be maintained at the same high level, exceeding the costs of operating power plants that will gain monopoly profits.

Therefore, generation capacities under the competitive market (with free prices) can expand only *at constant shortage of capacities* that is accompanied by high prices and superprofits of operating producers. Virtually, transition to the competitive market (price deregulation) *gives rise, in the long run, to the following dilemma*:

- Either new power plants will not be constructed at *low* prices of the wholesale market that correspond to costs of *operating* power plants, which will lead to electricity *deficit*.
- Or the *prices must rise* by 2–4 ¢/kWh (and even higher) with the associated *consequences* for the economy and population and with superprofits for the operating electricity producers.

None of the alternatives is acceptable and *this contradiction* characterizes the key property of the competitive electricity market in the long run, when the long-run production costs are costs of new power plants. This market property is caused by the properties of EPSs, primarily by their facility-by-facility expansion and also by a special financing mechanism for new power plants in the competitive market. The given property (contradiction) once again underlines imperfection of the electricity market. In fact, the competitive wholesale markets retain a market power (dominance) of electricity producers over consumers. They can *create deficit* and *soar prices* by *terminating* construction of new power plants. And the price barrier contributes to this, strengthening producers' incentives to stop construction.

The dilemma (contradiction or flaw) of the long-run electricity market can be resolved or eliminated *only by the state price regulation* and centralized planning of EPS expansion. *Higher prices* that are required for investment recovery *must be attributed only to new power plants*, not to operating ones. This is just the case in the single-buyer market (in the regulated monopoly this is implemented even simpler). Without price regulation, this shortcoming will certainly manifest itself, which is indicated by the experience of Brazil and Chile (see Chap. 7).

Figure 6.3 also shows that construction of new power plants of various types requires different prices of the wholesale market. They are the lowest for gas-fired CPPs with CCIs (4.93 ¢/kWh). Exactly such (and only such) power plants were constructed under the competitive market in Western Europe, Australia, and North and South America. The formed prices of natural gas, equipment of CCIs, and electricity generated by *operating* NPPs and CPPs on coal there allowed the investments in new power plants with CCIs to be recouped.

By the found estimates (in unpublished sources) construction of new CPPs on gas with CCIs required the wholesale prices of about 3.8 ¢/kWh (20 £/MWh) in Great Britain in the late twentieth century and 3 ¢/kWh (40 Australian $/MWh) in Australia at the beginning of the current century. Despite essential fluctuations, the actual wholesale prices in these countries on the average exceeded the indicated values, which encouraged construction of such power plants. In the second half of the 1990s Great Britain saw even construction boom of CCIs on gas.

Very high prices are needed for attraction of investments in new NPPs, CPPs on coal, and the more so in HPPs. Countries with competitive market virtually stopped constructing these types of power plants. And in countries where there are no cheap natural gas resources, and construction of such capital-intensive power plants is indispensable, transition to the competitive market proved to be simply impossible, since it would lead to extremely sharp rise of electricity prices. As will be seen from a review of markets in Chap. 7, this fact contributed to retention of regulated markets in both developed and developing countries, where construction of NPPs, HPPs, CPPs on coal, or power plants on unconventional renewable energy sources is necessary.

Analysis of Fig. 6.3 leads to one more conclusion: In the competitive market new power plants will be built in order of decreasing their financial effectiveness for an investor, i.e., based on the criterion of the earliest investment recovery, at first CPPs on gas with CCIs, then NPPs and so on. It means that without state regulation formation of the generation capacity structure *will not be really optimal*, as it should be by the criterion of *minimum total expenditures* for EPS expansion and operation. Eventually this will result in additional increase in costs for electricity production and its prices. The optimal structure of generation capacities can be ensured only at centralized designing and planning of EPS expansion, which is possible at market organization according to Models 1 and 2.

In conclusion, it should be noted that in some competitive markets (for example, the PJM market in the USA) attempts are made to overcome the considered flaw (price barrier) by organizing a so-called capacity market (it is also foreseen in the concept of the Russian NOREM). Theoretically, the concepts of capacity markets in combination with spot markets have been poorly thought through, and are subjected to criticism and revision. As a rule, only power plants with gas turbine installations (GTIs) and combined-cycle installations (CCIs) on natural gas are supposed to participate in trade in the capacity market, though they may happen to be insufficient and other types of power plants may be needed. It might be expected that the capacity markets are likely to fail as the spot markets. Their organization can be treated as another attempt of electricity producers to avoid regulation by all possible measures.

Box 17 Price Barrier in the Long Run and Its Consequences

1. In the *competitive* wholesale market higher costs of *new* power plants to be considered as long-run costs of NPPs and PGCs result in the *price barrier* for their construction in the *long run*. This barrier was illustrated on the example of EUPS of Russia for the period beyond 2010.
2. The following dilemma (contradiction) occurs:
   (a) Either under the wholesale market prices that correspond to the costs of *operating* power plants, new power plants will not be constructed and this will result in capacity *shortage* and deficit in the market.
   (b) Or the prices *should rise* by 2–4 ¢/kWh (and even higher) causing *damage* to the economy and population and bringing *monopoly* profits for *operating* producers.

"Deficit" is taken to mean the market state, when there is capacity shortage and equilibrium is established on *the vertical* range of the producers' supply curve, the price *exceeds* production costs and all producers gain a *monopoly* profit.

3. The indicated contradiction (a flaw of electricity market in the long run) can be resolved only by the *state regulation*. High prices that are necessary to recover investments should be permissible to NPPs only. Without regulation generation capacities in EPSs can expand only under *permanent shortage* of capacities (and electricity), and this will be accompanied by high prices and extra profits of operating producers. This is another proof of electricity market imperfection, in this case in the context of a long-term period.
4. The lowest prices in the wholesale market are necessary to construct *new* power plants with gas-fired CCIs. Such plants were constructed only in the countries that introduced a competitive market. They ceased to construct capital-intensive NPPs, HPPs, and CPPs on coal. However, in the countries that have no cheap natural gas resources and construction of such capital-intensive power plants is necessary, transition to a competitive market is simply impossible, since this will make electricity prices soar.
5. To overcome the above flaw the "capacity markets" are organized in some competitive markets (for example, the PJM market in the USA). The concepts of such markets are theoretically poorly thought through and are likely to fail similarly as spot markets. Their organization can be viewed as another attempt of electricity producers to avoid regulation.

6.5 Substantiation of the Efficiency of Intersystem and Interstate Electric Ties Under Different Models of Market Organization

This section is based on the long-term studies carried out by the author and his colleagues in the field of efficiency estimation of interstate electric ties in Northeast Asia. The results are summarized in the monograph [24]. The studies revealed the considerable impact the market organization models in the countries to be connected can have on the financing mechanisms for interstate tie construction and, hence, on the estimation technique of their efficiency. While under regulated markets (Models 1 and 2) the interstate tie efficiency estimation and financing do not pose any problems, under competitive markets (Models 3 and 4) serious *investment* problems arise which can result in rejection of the interstate tie construction even if it is *economically* efficient. These problems are directly related to *realization* of the power system interconnection effects considered in Sect. 2.2.

With some conditionality we can distinguish three interrelated problems of substantiating interstate ties under *competitive* market:

1. Electricity *export* is unprofitable for *consumers* in exporting country and *producers* in importing one.
2. It is hardly possible to realize the *capacity* effect of interconnecting power systems.
3. It is quite complicated to substantiate the *financial* (commercial) efficiency of interstate ties for potential investors.

Similar problems arise under competitive market in substantiating the efficiency of intersystem electric ties within one country. Therefore, both interstate and intersystem electric ties (ISETs) will be considered and commented upon below.

It should be noted that the above problems are very seldom mentioned and discussed in publications. This is why the author indicated them specially in Sect. 4.3 among the flaws of competitive markets and has dedicated this paragraph to them. These problems caused a dramatic decrease (or even cessation) of network construction including interstate ties in the countries that had passed to the competitive market. The situation is rather similar to the expansion of generation capacities in EPSs.

6.5.1 The Situation Under Regulated and Competitive Electricity Markets

In order to substantiate the ISET construction it is necessary to determine their economic and financial efficiency.

Economic efficiency is estimated by comparing two variants of expansion and operation of EPSs to be interconnected: with construction of ISET and without it. We will call them the variants of joint and separate EPS operation. Here different effects are taken into account: energy, environmental, social, and probably some others. These effects should be expressed in a monetary form (in rubles, dollars, etc.) or taken into account as constraints in addition to the economic estimation. Simultaneously, it is necessary to determine an expedient transfer capability of ISET, its design, and the required costs (capital and operating). While calculating the effects we have to apply mathematical models including those of power system expansion and appropriately compare lump sum investments and annual costs.

If the total economic (in money terms) expenditures, including costs of ISET, in the joint operation variant turn out to be lower than in the separate operation variant, the ISET construction is economically efficient. The economic efficiency is a necessary but not sufficient condition for ISET construction.

To implement the ISET project it is also necessary to show its *financial* (commercial) efficiency for *all* participants of the project (countries, companies, investors, etc.). Only in this case the agreement on its construction can be reached.

The process of financial efficiency estimation is also rather complex [72–74]. The main idea is to show that the revenues of each participant from the project implementation will exceed the costs borne. This was already discussed in Sect. 6.1 as applied to the estimation of financial efficiency of a new power plant for its investor. For the ISET project it will be even more complicated since there are several participants and it is necessary to estimate financial efficiency for each of them.

The estimation of ISET economic efficiency does not depend on the model of electricity market organization. The variants of joint and separate EPS operation should be compared in any case. The market model will affect the financing sources for ISET, the composition of project participants, and estimation of its *financial* efficiency for each participant. Let us consider it in more detail.

Figure 6.4 shows a general scheme of interaction among companies under market Models 1 and 2. These models are considered together since they have much in common.

The main similarity is that there will be single power companies at the ends of ISET: a VIC under Model 1 and a Purchasing Agency under Model 2. In principle, the same situation will arise in the case (which is not shown in the figure) with a VIC at one end of ISET and a Purchasing Agency at the other end. If ISET is economically efficient, these single companies gain a real benefit from its construction (EPS connection), for example by saving capital investments into new power plants. For a VIC this benefit and saving are quite obvious. A Purchasing Agency can also realize this effect by signing an agreement to jointly construct ISET and receive power from the neighboring system instead of concluding contracts with PGCs or NPPs for construction of new power plants on its territory.

Hence, if ISET is economically efficient, VICs and purchasing agencies *can finance* its construction by the money saved through the agreements to share the investments in the ISET and to exchange power and electricity. The total expenditures and tariffs for final consumers in both companies will be *lower* than under their separate operation (without ISET). Therefore, regulating bodies of both companies

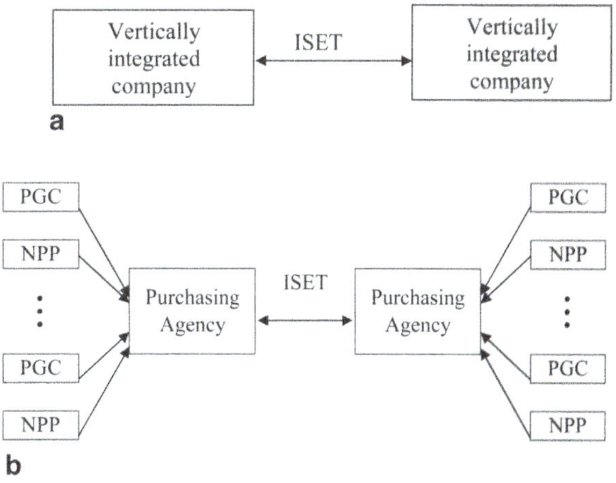

Fig. 6.4 Construction of ISET under **a** regulated monopoly, **b** single-buyer market model

harmonize the ISET construction and maintenance costs by including them in the electricity tariffs.

The financial efficiency estimation of ISET under market Models 1 and 2 is also relatively simple. In this case there will be two project participants—one from each country. It is also possible to create a joint venture (a subsidiary) for construction and operation of ISET. However, the conditions of its financial and economic activity are totally determined by the agreement achieved between the main participants. The financial efficiency of ISET for each country can be estimated by the generally accepted technique [72–74]. Naturally, it will be necessary to conduct negotiations on the prices of electricity to be exported, the distribution of ISET costs, payment for its use, etc., in order to ensure the financial efficiency of ISET for each country or EPS (for more details, see [24]).

It should be noted that the currently existing interstate ties and power pools in Western Europe and North America were created in the twentieth century during the era of regulated natural monopolies when there were no difficulties in financing the interstate ties.

Figure 6.5 shows the situation where the ISET to be substantiated connects the systems or countries with *competitive electricity markets*. Naturally, ISET penetrates to the wholesale markets where several independent power generation and sales companies (PGCs and SCs) compete with one another. ISET should become a participant in both wholesale markets. In doing so, the reversible ISETs will alternately play the role of electricity sellers or buyers. Such a situation will be typical of both Models 3 and 4 considered in Sect. 4.1. Therefore, further we will simply speak of the competitive market.

The conditions for substantiation and construction of ISET under competitive market will be totally different from Models 1 and 2 considered above:

- Now instead of single companies, the wholesale markets with many PGCs and SCs (without regard for large consumers that can directly enter the wholesale market avoiding SCs) appear at each end of ISET. It becomes unclear *what* companies (and how) will really benefit from construction of ISET and *finance its construction*, i.e. the participants of ISET project and its investors become uncertain.

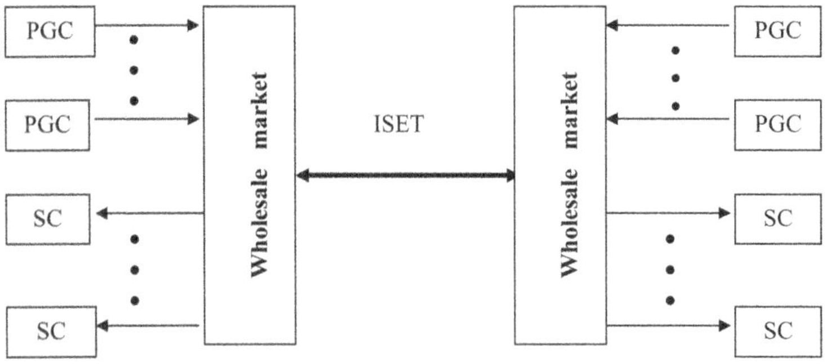

Fig. 6.5 ISET under competitive electricity market

- Instead of *negotiated prices* (tariffs) of electricity transmitted by ISET that were established by the parties concerned for a long period (10–20 years or even more) now in the wholesale markets the *equilibrium prices* will form according to supply and demand. These prices are characterized by high uncertainty and can hardly be practically forecasted for a long run.
- The wholesale market of each system or country has certain participation and trade rules; with ISET construction there will be a need to appropriately coordinate (match) these rules.
- Under the competitive market holding electricity generation, transportation (distribution), and selling within one company are forbidden. However, ISETs (particularly a reversible one) perform all the three functions—they transmit electricity (as a network company), supply it to one of the wholesale markets (similar to PGC), and purchase it in another market (like SCs). It is obvious that it is necessary to establish a special status for ISET as a participant of both wholesale markets.

This creates problems and troubles to be illustrated in more detail.

6.5.2 *Benefits or Losses Due to Electricity Export*

For illustration we will consider electricity export via an interstate tie though the same situation will be observed with electricity transfer via an intersystem tie within one country from the areas where it is cheaper.

Figure 6.6 taken from [69] shows the shift of market equilibrium when electricity is exported from the country (area) with lower electricity prices. In the exporting country the demand curve shifts to the right (the demand increases) and in the importing country the supply curve shifts to the right (the supply rises). Hence, the

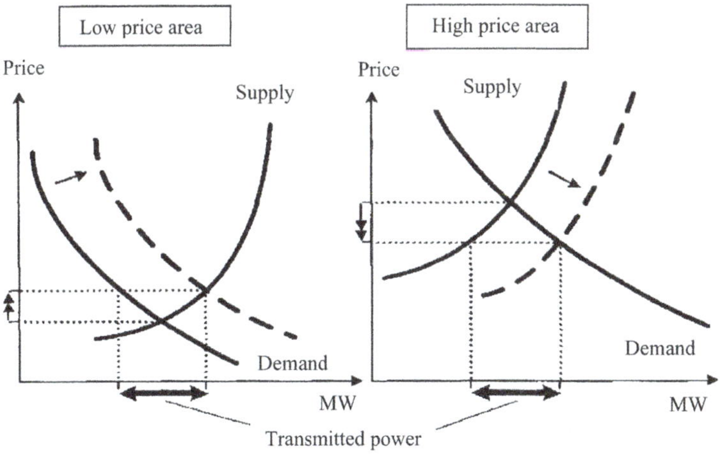

Fig. 6.6 Price changes during electricity export in the competitive markets

equilibrium price rises in the exporting country and declines in the importing country. This is a natural market process which however touches differently the interests of consumers and producers of the countries concerned.

Price rise in the exporting country will cause losses for consumers and benefits for electricity producers. On the other hand, in the importing country with *price decrease* consumers will benefit and electricity producers will suffer losses. Hence with electricity export under competitive market there will be market participants for whom export is not profitable. These are:

- Consumers in the exporting country
- Producers in the importing country

Naturally, they will oppose the export and construction of ISET.

The greatest benefit from export will be gained by the *producers in the exporting country*. This benefit will be double: from the *export itself* (rising sale at a higher price) and from the *increase in price* for the electricity sold in their own country. This will undoubtedly bring them superprofits.

Thus, with transition to the competitive electricity market *export ceases to be mutually beneficial*. ISET construction will undoubtedly face resistance as evidenced by practically absolute absence of new export transmission lines between the countries that have competitive markets.

Meanwhile, under the regulated markets (Models 1 and 2) export can be mutually beneficial. Internal electricity tariffs in the exporting country can decrease at the expense of revenues received from export. This is confirmed by intensive construction of ISETs, and formation of interstate power interconnections in the second half of the twentieth century before power industry deregulation.

As was already noted, a similar situation with prices in the competitive wholesale market will occur with construction of intersystem transmission lines connecting areas with different electricity prices within one country. There will also be contradictions between participants of the competitive market which will complicate the ISET construction.

6.5.3 Possibilities for Realization of Capacity Effect of Interconnecting Power Systems

We have considered the contradictions between the participants in the competitive market while constructing *export* ISETs with electricity transfer in one direction. Now let us analyze the case where the construction of ISET is *economically* efficient due to *reduction* of the required generation *capacities*, i.e., owing to the *capacity* effect. Such effects were briefly considered in Sect. 2.2. They can be achieved with EPSs interconnection owing to two factors: (1) a decrease in the emergency and maintenance capacity reserves and (2) a decrease in the coincident load maximum of consumers.

For simplification and better illustration we will consider only the second factor as applied to the interconnection of two power systems *within one country.* We will suppose that the ISET construction is economically efficient owing to the decrease in the coincident load maximum of the interconnected power system (IPS):

$$\Delta P_{\mathrm{IPS}}^{\max} = \left(P_1^{\max} + P_2^{\max}\right) - P_{\mathrm{IPS}}^{\max}, \tag{6.22}$$

where:

$P_1^{\max}$ and $P_2^{\max}$—are the annual maximum loads in power systems 1 and 2;
$P_{\mathrm{IPS}}^{\max}$—is the coincident annual maximum load in the IPS;
$\Delta P_{\mathrm{IPS}}^{\max}$—is the decrease in the coincident maximum load.

The demand for generation capacities in power systems decreases by the value of $\Delta P_{\mathrm{IPS}}^{\max}$ after their interconnection, i.e., with construction of ISET the commissioning of new power plants can be reduced. If the capital investments saved on construction of new power plants exceed the capital investments in ISET (K_{ISET}), the latter will be economically efficient:

$$K_{\mathrm{ISET}} < \Delta K_1^{gen} + \Delta K_2^{gen}, \tag{6.23}$$

where ΔK_1^{gen} and ΔK_2^{gen} are capital investments saved due to decrease in construction of new power plants in the first and second EPS.

For the sake of simplicity the other effects from EPS interconnection (the mentioned decrease in required reserves, fuel and cost savings, mitigation of environmental impacts, etc.) will not be dealt with here. Assume that inequality (6.23) is satisfied at ISET construction and it is sufficient to admit ISET as economically efficient.

Let us introduce one more assumption important for this case—*the wholesale prices in both EPSs are nearly the same*, i.e., ISET is efficient only owing to the capacity effect from EPS interconnection.

Note that the considered capacity effect (6.22) is particularly high when EPSs with different *seasons* of the annual maximum load are interconnected. Such a situation is typical, for example, of Northeast Asia, where in Russia, Democratic People's Republic of Korea (DPRK), and Mongolia the annual maximum load is in winter and, on the contrary, in Japan, Republic of Korea, and in most regions of China it is in summer. Specifically, the studies on the efficiency of the ISET "Russian Far East–DPRK–Republic of Korea" have shown that in 2020 decrease in the coincident maximum load of the three EPSs to be interconnected $\left(\Delta P_{\mathrm{IPS}}^{\max}\right)$ will amount to some 7.5 GW and the total saving of investments in generation capacities (ΔK^{gen}) will be \$13.4 billion at the ISET cost of \$1.5 billion [24]. In this case, ISET will be used for *reversible seasonal* power flows.

Let us continue discussing ISET within one country that is economically efficient by virtue of meeting inequality (6.23). When in EPSs to be interconnected the markets are organized based on Model 1 or 2 (Fig. 6.4), the single companies at the ISET ends will gain a real effect from its construction—decrease in commissioning new power plants and the appropriate investment saving. As already noted, they can finance ISET construction on parity basis owing to saving. It will be more advantageous for them to invest the corresponding share of funds in ISET and receive electricity from operating power plants of the neighboring EPS than make investments in construction of new power plants on their territory. The electricity tariffs in this case will be lower than at isolated operation of EPS and the regulatory bodies will give their consent to ISET construction and include the needed investments in the investment component of consumer tariffs instead of higher investments in new power plants.

Note that the ISETs realizing the capacity effect from EPS interconnection are similar to generation capacities of EPSs in many respects and are an alternative to the latter. If ISETs are economically efficient, they are the best means to meet growing consumer loads than construction of new power plants. As indicated in Sect. 2.4, the integral effect from creation of the UPS of the USSR was 1.5–2.5 times higher than the costs for development of the backbone network (see also [25]), and the total decrease in the coincident maximum load owing to interconnection of regional EPSs in IPS and IPSs in the UPS in December 1991 reached almost 14 GW [52]. Hence, in parallel with power plants ISETs supply power to individual EPSs and create economies of scale of the whole power interconnection. The problems of their financing and estimation of financial efficiency, therefore, have much in common with similar problems for power plants.

Suppose now that the annual maximum loads $\left(P_1^{\max} \text{ and } P_2^{\max}\right)$. are expected to *increase* in both EPSs and these increases are larger than half the decrease in the coincident maximum $\left(0.5\ \Delta P_{\text{IPS}}^{\max}\right)$. In this case both EPSs can gain an equal effect from ISET construction (investment savings owing to unnecessary construction of new capacities):

$$\Delta N_1^{gen} = \Delta N_2^{gen} = 0.5 \Delta P_{IPS}^{\max}. \tag{6.24}$$

To do this, the transfer capability of ISET (neglecting transmission losses) must be equal to the same value:

$$N_{\text{ISET}} = 0.5 \Delta P_{\text{IPS}}^{\max}, \tag{6.25}$$

and it will operate in a *reversible* mode, transmitting the indicated power alternately to EPS with the own annual maximum load. In this case 1 kW of transfer capability of ISET will save 2 kW of generation capacity (per 1 kW in EPS).

Relations (6.24) and (6.25) are optimal to provide complete realization of the capacity effect (6.22) at a minimum transfer capability of ISET. This transfer capability will be fully used in both directions *during the hours of annual maximum load* of the appropriate systems. Meanwhile, the total duration of annual maximums of EPS is relatively short and for the remaining time of the year there is no need to transmit power via ISET in order to implement the considered capacity effect. Hence, *electricity exchange* (double-sided) by such ISET *may be rather small.* Despite this fact the ISET construction is economically sound, provided condition (6.23) is satisfied. Therefore, as stated above, the *regulated companies*

(Fig. 6.4) will finance its construction and own it jointly at one or other organizational form.

In the environment of the *competitive* wholesale market (Fig. 6.5) the situation with the implementation of the capacity effect (6.22) will be much complicated. Because of separation of the spheres of electricity generation and transmission the generation capacities will belong to numerous PGCs in both EPSs and the main grids including ISETs to the national network company (NNC). The latter is regulated by the State and exists owing to the transmission fee for using networks. The fee also includes investments in network expansion. However, the need (effectiveness) for construction of every new transmission line should be substantiated and coordinated with a regulatory body. Construction of a new ISET will also require special substantiation. Virtually it implies substantiation of *financial* efficiency of ISET, i.e., indication of investment sources and demonstration of possibilities for investment return (payback). This gives rise to uncertainties, difficulties, and problems.

Consider at first potential investment *sources* for ISET construction. Obviously they cannot include funds of PGCs, financing only expansion of their generation capacities. The same refers to ISET construction by the SCs. In principle, it can be done by a private (external) investor; however, it is even more difficult to substantiate for him efficiency of the *reversible* ISET than for NNC. The transmission fees received by NNC allow only gradual *payback* of the investments, but they cannot include large capital investments in the period of ISET construction.

Bank *credits* to be returned later seem to be the sole real source of ISET construction under competitive market. (And this is instead of the saving of investments in construction of new power plants under regulated electricity markets.)

To *get* a credit (and to coordinate this with a regulatory body) NNC should show how (owing to what revenues) it can be repaid. *Charge for ISET use* (transmission of power and electricity via it) that is *collected from PGCs and SCs*, which will trade in electricity between EPSs to be interconnected, is a natural source of such revenues. NNC, therefore, must *elaborate a business plan* (with corresponding financial flows) that will demonstrate repayment of investments (credit) in ISET. In accordance with the accepted techniques [72–74] the business plan is drawn up for a *long* period of time covering the terms of ISET construction and service. For this plan to be realistic (guaranteed) the NNC's tasks are to:

1. *Attract PGCs and SCs for plan elaboration* from both EPSs to be interconnected that will use ISET. NNC must conclude the *long-term contracts* (for 10–15 years needed for credit repayment) with them that would guarantee ISET use and proper payment. Such contracts will make it possible to guarantee credit repayment and to convince a bank and regulatory body.
2. *Establish such charge for ISET use* that would satisfy both PGC and SC, on the one hand, and cover credit repayment with the volumes of electricity and power transfer settled in the contracts, on the other hand.

To analyze the possibilities for meeting these conditions it should be reminded that here a *reversible* ISET intended for realization of the *capacity* effect owing to EPS interconnection is dealt with and such ISET will be used only for a short period per year *during hours of annual maximum loads* of EPSs to be interconnected. Indicate also the assumption made on *roughly equal prices* in the wholesale markets of both EPSs and that ISET is *economically* efficient owing to the capacity effect only, i.e., satisfaction of inequality (6.23).

Trade in electricity through ISET involves sale of electricity by the generation companies of one EPS to the sales company of the other EPS. For the reversible ISET, aimed at realization of the capacity effect from the EPS interconnection, the electricity will be traded in both directions, i.e., the pairs "PGC–SC" should be formed for each direction. These pairs would also conclude long-term contracts with each other.

Besides, at small volumes of transmitted electricity and power the charge for ISET use must be *high enough* to pay back investments in it (to repay credit). And this charge will be added to the expenditures of PGC and SC trading through ISET.

With approximately equal wholesale electricity prices in both EPSs and the additional charge for ISET use, there *seem to be no incentives* (reasons) for concluding contracts between PGCs and SCs of different systems. Theoretically it can be imagined that during the hours of the own maximum of each EPS the wholesale price there will be higher than during the same hours of the other EPS. However, it will be practically impossible to "detect" this difference when concluding long-term (for 10–15 years) contracts, in particular when the maximum loads of both systems coincide seasonally. And it is doubtful whether this difference exceeds the required charge for ISET use.

Thus, under competitive market:

- A reversible ISET can be constructed by NNC only on the basis of bank credits.
- Capital investments (credits) in ISET should be repaid solely by the charges for its use by the PGCs and SCs of different EPSs. The charge will be very high because of short-term flows only during the hours of annual maximum loads.
- NNC must elaborate a business plan to substantiate financial efficiency of the ISET project. In this case PGCs and SCs from different EPSs should be involved (on the basis of the long-term contracts) in the project (and business plan). This will guarantee an actual use of ISET (and proper payment).
- Meanwhile, at almost equal wholesale prices in both EPSs (ISET is intended only for realization of the capacity effect) PGCs and SCs have no incentives to trade through ISET with an additional charge. NNC, therefore, will not be able to get them involved in the ISET project.

Consequently, NNC *will not be able to come to an agreement with PGCs and SCs* about a reversible power transmission through ISET and substantiate its *financial* efficiency. Therefore, *construction of a reversible ISET* to realize the capacity effect from EPS interconnection *becomes virtually impossible under competitive market.*

A similar and even more intricate situation occurs in the competitive market with the *reversible interstate electric ties* that realize the capacity effect owing to interconnection of EPSs of different countries. In this case special intergovernmental agreements on mutual supplies of power, contracts between NNCs of participating countries, etc., are required [24].

6.5.4 Difficulties in Substantiating Financial Efficiency of Interstate Ties Under Competitive Market

The considered features of electricity export and possibilities for realization of the capacity effect from EPS interconnection give a general idea about difficulties in substantiating the *financial* efficiency of ISET in the competitive market. It is typical of both export (one-sided) and reversible transmission lines.

These problems in the context of *interstate electric ties* are dealt with in great detail in [24]. Here we note only their main aspects.

Similar to the situation with generation capacities the transition from the regulated markets to the competitive one leads to change in sources and mechanisms of ISET financing. In addition to demonstration of economic efficiency it is needed to substantiate financial efficiency for investors and all the rest of the participants in the ISET project. Some difficulties emerging in this case have been already considered or mentioned above. The main problems described in [24] consist of the following:

1. Settlement of the question about the investor (and owner) of ISET: a private (external) investor or national network companies of the countries to be interconnected. For the private investor it is an extremely hard task to substantiate financial efficiency of ISET. As to the national network companies they can finance ISET construction only through the bank credits.
2. Necessity to attract PGCs and SCs of interconnected countries as participants of the ISET projects. In this case several pairs of "PGC–SC" should be formed with allocation of the total transfer capability of ISET and the total volumes of electricity transmitted among these pairs.
3. Necessity to conclude long-term contracts among participants of the ISET project for a period no less than the time for investment (credit) repayment.
4. Establishment of such a charge for ISET use, at which the ISET project would be financially efficient for all its participants (investor or creditor, NNCs, PGCs, and SCs of the interconnected countries).

As shown in [24], in principle (theoretically) it seems possible to substantiate ISET efficiency for the competitive markets at its end. However, the indicated difficulties make such substantiation extremely sophisticated.

This is confirmed by only a few instances of constructing new ISETs between countries with a competitive market. And some of them (e.g. the export ISETs from France to Spain) have a country with the regulated electricity market at one of its ends.

Meanwhile, before the transition to a competitive market, ISETs were intensively constructed in Western Europe and North America. At present their construction is going on in Asia, Africa, and the Middle East between the countries with regulated electric power industry.

Box 18 Investment and Financial Efficiency of ISETs

1. To substantiate the construction of intersystem or interstate electric ties (ISETs) it is necessary to estimate their *economic and financial* efficiency independently.
   Economic efficiency is estimated by comparing all kinds of *expenditures* for two variants of expansion and operation for power systems to be interconnected: at their separate operation (without ISET) and at their joint operation (with ISET). Should the total expenditures, including the costs of ISET, in the variant of joint operation *be lower* than in the separate operation variant, the ISET is *economically* efficient. This estimation should be made in any case.
   The estimation of *financial* efficiency is necessary to show that the revenues of each participant from the ISET project implementation will *exceed the expenditures of the participant.* The financial efficiency of ISET *depends on the model* of electricity market organization: sources of financing, participants of the project, etc.
2. In regulated markets (Models 1 and 2) there will be single energy companies on the ends of ISET: a vertically integrated monopoly or a Purchasing Agency. These companies will *get the effects* from connecting their power systems and, if ISET is *economically* efficient, they may finance its construction on a parity basis. Estimation of the *financial* efficiency does not cause any problems.
3. If ISET connects the systems (or countries) with *competitive* markets, there will be wholesale markets on its ends, in which several independent PGCs and SCs compete with one another. It becomes unclear *what* companies (and how) will *really get the effect* of ISET construction and *finance* its construction. At the same time the uncertainty of prices in the wholesale markets and other problems occur.
4. One of the problems is *unprofitability* of electricity export for consumers in the exporting country and producers in the importing country, since in the exporting country the demand and prices increase and in the importing country the supply rises and prices decline. This will inevitably cause *opposition* and complicate the ISET construction. The electricity export is *no longer mutually beneficial* in the competitive market.
5. In the competitive market it is particularly difficult to substantiate the *reverse* ISET intended for implementation of *capacity effect* of EPS interconnection, in other words, a *decrease in demand for generation capacities* with construction of such transmission lines. This is explained by

separation of electricity generation and transportation businesses and *change in the financing mechanism* for ISET as compared to the regulated markets.

6. Construction of an interstate electric tie requires special intergovernmental agreements. Though *theoretically* their substantiation under *competitive* market seems possible, yet practically it becomes extremely complicated. This is confirmed by a sharp reduction of ISET construction between countries with competitive markets in the electric power industry.

Chapter 7
Worldwide Experience in Electric Power Industry Restructuring[1]

This chapter presents an analysis of certain conditions, goals, forms (models), and results of the electricity reform in various countries of the world. They are sufficiently different depending on the economic development, available energy resources, political structure, and other features of a country. Reforms in developed countries started in rather favorable conditions (large generator reserves, low demand rates, fair network development, etc.), and the final goal was to reduce electricity prices (tariffs) for the customers. In developing countries, the reform is caused as a rule by the electricity deficit, insufficient state investments, and other "growing pains." The depth of the reform varies even for different regions of large countries (USA, Canada, India) consisting of several states or provinces.

Still, the reform outcomes are common for the countries with preserved electricity price regulation (using market Models 1 and 2), and countries that got over to the competitive market (using Models 3 and 4). Such different results are noticed both in developed and in developing countries, in the same world regions, and even within one country. The US experience is of interest here, for there are states that have retained the regulated monopolies as well as states performing the electricity markets deregulation.

Taking into consideration the above-mentioned diversity of conditions and goals of reform and the commonality of its results, the reforming experience analysis is performed for the groups of countries that have retained electricity price regulation (Sect. 7.2), and those performing market deregulation (Sect. 7.3). A more detailed examination of the electricity reform is outlined as an exception in Sect. 7.1 for the USA and Canada, where there are both processes at the same time.

[1] A mutual paper with Dr. V. V. Khudyakov [22] was used in this chapter.

L. S. Belyaev, *Electricity Market Reforms*,
DOI 10.1007/978-1-4419-5612-5_7,

7.1 Power Industry Restructuring in the USA and Canada

It is often written and presented that the USA stands "in the front line" of the electric power industry reforming to a competitive market. Meanwhile, as of the end of 2005 (and minor changes have occurred since then) the majority (27) of states in the USA did not start deregulation of the market at all, retaining vertically integrated regulated companies (Model 1). Four states, including California, began reforming, but then stopped it, resuming regulation. Another three states have implemented partial restructuring, and only one-third of the states (17) transferred to a competitive electricity market.

A more impressing situation is in Canada, where only one province (Alberta) has shifted to the competitive market, and another one (Ontario) tried to do so, but was unsuccessful. Other provinces retain the regulated monopolies.

7.1.1 The Start of the Reform in the USA

Prior to the 1970s, the US electric power industry successfully developed in the form of regulated vertically integrated companies (VICs). Generation capacities were abundant, electricity was exchanged among VICs of different states, and three bulk interconnected power systems with back-to-back DC links between them and transmission lines to Canada and Mexico were formed. The electricity tariffs were essentially different over the country, but they gradually decreased.

The world energy crisis caused by a bounce in the oil prices in 1973 and 1979, more extensive use of natural gas and renewable energy sources (RES), and some other factors had challenged the improvement of the electric power industry management.

Under the Regulatory Act of 1978 (PURPA), independent power producers (IPPs), mostly industrial cogeneration power plants (CGPPs) and RES plants, began selling electricity to VICs through long-term contracts [6]. The later Energy Policy Act of 1992 (EPA) granted additional terms for the development of IPPs and expansion of the wholesale trade between the states with the retention of VICs.

The real transition to the competitive market commenced in 1998 in three states (Massachusetts, Rhode Island, and California) after extensive debates and several resolutions of the Federal Energy Regulatory Commission (FERC). By the year 2000, about ten states with the most expensive electricity joined them, and approximately ten other states signified their intention to a similar reform [6]. In the remaining states there was a broad opposition to deregulation.

Sufficient influence on transition to the competitive market rendered the position of the large industrial consumers who bought electricity from the regulated VICs at high retail prices, based on the misunderstanding. The point was that the electricity markets between VICs of one or several states were called "wholesale." The prices at these markets were low because practically all VICs had large power reserves and traded in the "surplus" at the *variable* (fuel) costs. At the same time, the regulatory authorities naturally included the *fixed* power plants costs into the retail prices, in particular the repayment of credits borrowed for their construction. In other words, the fixed costs of generation (with the investment component) were absent in the so-called wholesale prices and were included in the retail prices. This circumstance created a great difference between the "wholesale" and retail prices, and misled the customers—an illusion appeared that elimination of the regulation would result in decrease in retail prices [6].

7.1.2 Energy Crisis in California[2]

The *wholesale* competitive market (Model 3) was implemented in California in April 1998. Gas power plants owned by the three VICs and comprising about a half of the power generation of the State had been sold to the five independent power-generating companies (PGCs) in advance. Though the nuclear and coal power plants had been launched for sale, this capacity continued to be controlled by the former VICs at the start of the crisis. VICs were restructured (with the transmission networks assigned) and transformed into large distribution-sales companies (DSCs), which supplied the customers under warranty at the *regulated retail prices*.[3] Some small IPPs remained. California imported large quantities of electricity from the neighboring states at the periods of high demand.

The California Independent System Operator (CAISO), the day-ahead market, the balancing hour-ahead market, ancillary services markets, etc., were created. California's restructuring and competition rules required that all electrical energy trade should be provided through the spot day-ahead market.

It is important to mention that the market conception has foreseen one more "nonstandard" wholesale market player (besides the direct power producers and DSCs)—the "reseller," who could make contracts with both the producers and buyers of electricity. One of such "resellers" got the notorious Enron Corporation.

There was a strong belief that the wholesale prices should decline due to the competition effect. The belief was so strong that the retail prices were frozen for 4–5 years at the level of about 6 ¢/kWh. This rather high price was established yet in 1996 to compensate for the "stranded costs." The expectation was that generation expenses would be rather lower than this figure (around 3 ¢/kWh), and this will help to fast realize such compensation, after that retail prices might be defrosted (and should decline).

The market functioned normally for two years (up to April 2000). The spot market prices fluctuated rather high, but their average value was kept at the expected

[2] The crisis is described mostly according to the unpublished paper: "A Quantitative Analysis of Pricing Behavior in California's Wholesale Electricity Market During Summer 2000: The Final Word." By Paul Joskow and Edward Kahn. February 4, 2001.

[3] These three large companies later suffered the most detriment from the crisis or even went bankrupt.

level of 3 ¢/kWh for 1998–1999. Meanwhile, not one new power plant was commissioned during this period, and the demand increased. The summer peak load in the state area reached about 43 GW in 1999 (demand fell to less than 20 GW during some off-peak periods in spring and autumn).

The summer in 2000 was extremely hot, which resulted in energy deficit appearing in the California networks (and in neighboring states as well). Customers' demand drastically increased, hydropower plants (HPPs) suffered low water rate, electricity import decreased. Wholesale day-ahead hourly prices in July 2000 increased on average to 13.2 ¢/kWh, and in August and December to 17.5 and 38.5 ¢/kWh respectively. The price of spinning reserve of generators jumped from 1 ¢/kWh in the beginning of 1998 to 75 ¢/kWh in June 2000, and to 150 ¢/kWh in December 2000. The same high prices remained in the first part of 2001. At the same time the natural gas price increased 2.5 times, and tradable permits for NOx emissions increased as well.

This resulted in an energy crisis burst in California in 2000–2001, with a series of blackouts and deliberate tripping of consumers. While the retail prices were fixed, the consumers did not react to the rise of the wholesale prices. DSCs incurred tremendous losses and went bankrupt. At the same time, PGCs, and "resellers," especially Enron Corporation, made profits up to 350%. The wholesale market actually ceased to operate in January 2001. The curtailment of the spot *price cap* by the regulatory authorities from 75 ¢/kWh to 50 ¢/kWh in July and to 25 ¢/kWh in August 2000 did not help. The State Government and FERC alleged that the market deregulation in California had failed; they fined some sellers and restored the wholesale price regulation.

Numerous subsequent investigations showed that an unprecedented rise in the wholesale prices during the crisis could not be explained by the objective conditions formed (augmented electricity demand, low water flow, import decline, natural gas price rise, and tradable permits for NOx emissions increase). There appeared actual use of market power—manipulations of the independent PGCs, and "resellers," including the deliberate lockout of generators and transmission lines to form deficit. Some of these manipulations were discovered later, after the bankruptcy of Enron, in the course of court trials [51]. Nuclear power plants (NPPs) and condensing power plants (CPPs), which were kept by the three bankrupted DSCs (former VICs), were naturally operated to the highest possible extent. Some natural gas price manipulations were discovered as well. The behavior of the independent producers and "resellers" during the California crisis confirms the insolvency of the spot electricity markets, which was considered in Sect. 5.2.

7.1.3 Reform Outcome

The energy crisis in California sufficiently slowed down the reforming process. Nine states that had planned or started deregulation rejected or canceled it. The only state that joined reforming again after the California crisis was Texas.

For the first time in 15 years the average retail electricity prices in the USA increased for industrial consumers in 2000 and for residents in 2001. In August 2002, FERC proposed a Standard Market Design (SMD), which, however, was declined by majority of the states. They believed that such reforming of the electric power industry did not meet their consumers' interests. As a result, within a year (in August 2003), FERC issued a new resolution (White Paper) that granted a longer time and greater freedom to the states in the implementation of reforms. Thereafter, the competitive wholesale and retail markets were further developed, mainly in the Northeastern states and also in several states of the Midwest and Texas.

The blackout of August 14, 2003 (the most severe in the history), in Northeastern USA, which had transferred to the competitive market and the bordering provinces of Canada, was the second indicative event (after the Californian crisis). It involved disconnection of 61.8 GW of load and affected 50 million consumers. In some areas of the USA, electricity supply was restored only after 4 days, and in some provinces in Ontario it took a week. The detriment caused by this emergency in the USA is estimated to be from US $4 to US $10 billion, and in Canada—about CAN $2.3 billion [76].

After this blackout FERC and the North America Reliability Councils stiffened the national standards on maintenance of reserves, frequency, and voltage, made the reliability standards mandatory, and also set control of their observance. At the same time this blackout contained the striving for deregulation and restrained the process of power industry reform in the USA.

In 3 years, on April 17, 2006, one more blackout with the forced tripping of load occurred in Texas which had also transferred to the competitive market [77]. Though in [76, 77], both the blackouts are not linked directly with transition to the competitive market, however, their occurrence in the power systems with such a market is a reality. Reliability decrease in the process of power industry deregulation can be explained in many ways: disintegration of the single VIC into dozens of "ill-assorted" companies with their own and, as a rule, contradictory interests; advent of congestion management problem; limitation of functions of the System Operator due to the activity of the Trading System Administrator; difficulties associated with the investments in new power plants and maintenance of required capacity reserves, etc.

The final goal of the power industry deregulation was *to decrease electricity prices (tariffs)*. This has recently raised the question of to what extent this goal was achieved. The paper [2] reveals the difficulties in answering this question, since prices change in all the states (both deregulated and not deregulated) in response to many factors—inflation, change in fuel prices, etc. However, despite these difficulties the price dynamics in different states has been compared in several papers.

The most comprehensive analysis of prices for industrial consumers for the period 1990–2003 is presented in [7]. The average annual rates of price change (decrease or increase) for the periods before the beginning of the power industry reform and after it are taken as the basic indicator. For the states implementing deregulation, the first period was taken individually from 1990 to the actual start of reforming in the state. For the states retaining or restoring regulation, this period was assumed to be from 1990 to March 1998. The second period covered the years after the beginning of deregulation (for the states implementing it) or 2001–2003 for the states that did not carry out or stopped the reforms. The results of the analysis have shown that, on the whole, the average annual rates of electricity price variation for groups of the continental states (except for Alaska and Hawaii) were as follows:

- In the states with the competitive market the rates changed from 0.3% in the period before the beginning of reform to 1.7% in the subsequent period, i.e., an increase of 1.4%.
- In the states with regulation the rates changed from –0.7% in the first period to 0.1% in the second period, i.e., the rates increased by 0.8%. This value is much lower than in the previous group of states.

A similar analysis of prices is described in [11] for the period from April 2005 to April 2006, when they rose virtually in the whole country, primarily because of increase in natural gas prices. The average electricity prices in the USA increased during this year by 10.9%. The highest increase was observed in the states that had introduced deregulation; in particular, in Texas it reached 46.4%.

The latest (known to the author) studies of electricity prices for the period from January 2003 to May 2007 are presented in [12]. The paper compared retail prices for the group of states that implemented deregulation, and states that retained regulation, and also prices in the neighboring states Texas and Louisiana (with regulated prices). The natural gas prices, whose rise slightly differs for the indicated groups of states, and the difference in prices excluding the fuel component were also analyzed. In short, the results of these studies are as follows:

- The difference in electricity prices between the groups of states varies from 1 ¢/kWh to 2.3 ¢/kWh for the benefit of the states retaining regulation, with the visible trend toward its increase in the recent years.
- The difference in prices without fuel costs has proved to be even greater—from 1 ¢/kWh to 2.7 ¢/kWh, also with the trend toward its increase.
- Since 2005 the prices in Texas have been regularly 1–2.5 ¢/kWh higher than those in Louisiana, the difference also tending toward increase.

The authors of [12] logically explain higher prices in the states with deregulated power industry by the fact that the producer's surplus is pocketed by PGCs. Meanwhile, in price regulation this "surplus" is withdrawn to the advantage of electricity consumers, thus decreasing retail prices.

The described studies reveal that instead of electricity price decreasing for final consumers, the reforming (deregulation) in the USA, on the contrary, brought about a price rise. Even if the competition in the wholesale and retail markets resulted in some benefit concerning decrease in production costs, the whole benefit was derived by generating companies (together with the producer's surplus). The consumers simply incurred losses due to the price rise. It is not strange that the majority of states retain regulation, taking care of competitiveness of their economy and the interests of their population.

Currently, the retail electricity prices rise more and more in the states with competitive market compared to those with regulated monopolies. Seventeen states that implemented deregulation have taken an increasing number of new measures to overcome difficulties that have arisen: introducing the capacity market to guarantee timely development of generation capacities, markets of derivatives (futures, options), etc. The electricity markets in these states became very sophisticated. And the experience shows that basically electricity producers benefit from deregulation.

7.1.4 Expansion of Generation Capacities

In the last two decades, the generation capacities in the USA have been developing in cycles. Decrease in the construction of new power plants in the 1990s was replaced by the "boom" in the first years of the current century. Capacity reserves (average for the country) decreased from 35% in 1985, to 15% in 2000, and increased to 30% again in 2004 [78, 79]. In the most recent years construction has declined anew.

Prior to the 1980s all types of power plants, including HPPs and NPPs, were constructed, as a rule, with overinvestment, i.e., excessive construction with formation of unreasonably large capacity reserves. By the way, the regulated vertically integrated monopolies were believed to be "guilty" of it and this fact was an argument for the transition to the competitive market.

Decline in construction in the 1990s can be explained by two principal reasons: first, the more accurate planning (and regulation) of power system expansion in the states with VICs, and second, the high uncertainty and risks for the investors in the states that began or planned to move to the competitive market. On the whole, reduction of reserves to the normal level should be treated as a positive factor.

The beginning of the current century is characterized by intensive construction of combined-cycle power plants (CCPPs) on natural gas in almost all states. It was favored by the growing power consumption and low gas prices (about $70/tce). Construction of CCPPs with low capital investments and high efficiency was financially attractive under those conditions at the wholesale prices corresponding to costs of the operating nuclear and coal-fired power plants. A similar "boom" in the construction of CCPPs occurred in England in the 1990s. Meanwhile, as far as the author knows, no new NPP or HPP has been constructed in the last decade in the USA, and no coal-fired condensing power plant has been constructed in the states that transferred to the competitive market. The construction of capital-intensive power plants became unprofitable as the investments would not return, as was shown in Chap. 6.

On the one hand, the possibility of CCPP construction is a favorable factor, since they contribute to expansion of generation capacities and elimination of power shortage. On the other hand, however, there are certain adverse effects. CCPPs were built primarily by independent producers "at their risk" without proper consideration of future conditions and, naturally, without coordination with one another. These facts resulted in the new "overinvestment" that reduced the capacity factor (utilization of capacities during a year) of the new power plants, and decreased the investment payback against the expectations.

A sharp increase in the natural gas prices—2.5 times by 2004–2005 (up to $170/tce) and even higher in 2006–2007—proved to be the gravest "blow" [12, 78]. Many generating companies were found to be in the heaviest financial state. The cost of shares in some of them fell from $40–60/share in May 2001 to $3–6/share in March 2003 [6]. In 2004, the net revenue for the majority of companies was less than 50% of the expected one [79]. About 125 GW of new capacities (projects) planned before 2001 were canceled or deferred for an indefinite time [6]. Thus, at present another deceleration of generation capacity development starts up.

7.1.5 Canada

In the province of *Ontario* the competitive electricity market failed almost as soon as it was introduced in 2002 [2, 80]. In 6 months the prices soared and their regulation was restored. The Government made new attempts to implement the competitive wholesale and retail markets; however, they could hardly be successful due to the growing shortage of generation capacities [80].

The market in the province of *Alberta* was estimated in [4] as an unsuccessful one because of the following: the manifestation of market power, mostly when congestion occurred in the electric network lines, insufficient commissioning of generation capacities, great volatility and a general growth of electricity prices at a weak response of consumers to their changes, etc. Supposedly, since that time some measures were taken to improve the market, but unfortunately the author has no information on its current state.

It should be noted that the province of Alberta is the only one in Canada that introduced a competitive market. The remaining provinces in Canada retain regulated vertically integrated monopolies, similar to most states in the USA.

Box 19 Results of Electric Power Industry Restructuring in the USA and Canada

1. Owing to a considerable autonomy of states and provinces and an active opposition to deregulation, most of the states in the USA and almost all provinces in Canada retained regulated vertically integrated power companies with the open access for the IPPs. No crises and abnormal phenomena in the development and operation of electric power systems (EPS) have been observed in these states and provinces. They have definitely improved the regulation methods and procedures.

2. The state of California and the province of Ontario encountered crises shortly after organization of the wholesale electricity market. These crises were accompanied by the deficit, a skyrocketing rise of prices, emergencies, and disconnections of consumers. Afterwards the price regulation was restored there.
3. In the states of the USA and in the province of Alberta that transferred to the competitive market, its drawbacks have appeared quite obviously:
   a. There was an essential (by 1–2.5 ¢/kWh) increase in prices as compared to the states with regulation retained.
   b. Construction of capital-intensive HPPs, NPPs, and CPPs on coal stopped; the natural gas-fired CCPPs were the only ones constructed during the periods when natural gas was cheap.
   c. System blackouts occurred in the northeast of the USA in 2003 and in the state of Texas in 2006.
   d. The investments in electric networks decreased, the line congestion problems arose, etc.
4. The attempts to eliminate or mitigate these flaws lead to the permanent complication of competitive markets: implement a payment for the capacity, creation of markets for ancillary services, capacity markets, markets of derivatives, etc.
5. In the early twenty-first century there was a "boom" of CCPP construction in the USA, and hence capacity reserves increased up to 30% in 2004. This is indicative of the possibility "to overinvest" under the competitive market (previously this overinvestment was considered to be a flaw intrinsic to regulated monopolies).
6. No state in the USA or province in Canada made a transition from regulated monopoly to the single-buyer market. This can be explained by the low demand growth, and primarily private ownership of the monopoly companies.

7.2 Positive Examples of Markets with Regulated Prices

In the countries with the state control of the power industry that covers regulation of electricity prices (tariffs) (Models 1 and 2), there are no shortcomings and consequences inherent in the competitive markets (Models 3 and 4). China, India, South Korea, France, and Japan present examples of such countries. The power industry of these countries is developing and operating successfully despite the problems caused by the rapid growth of power consumption or poor provision of own energy resources.

7.2.1 China

The power industry reforming was launched in 1985 [42, 81] when the State Council of China adopted the resolution for encouragement of nongovernmental investments in the energy sector to eliminate the existing power deficit. The resolution facilitated the signing of long-term contracts with domestic and foreign private investors by federal or provincial governmental bodies. The electricity prices in the contracts were fixed so that they guarantee the annual return of investment (usually more than 15%). A great number of IPPs appeared due to these measures, and shortage of generation capacities was eliminated by 1997 in almost the entire country.

In 1997–1998 the functions and responsibilities of the State were separated from the direct economic activity of energy enterprises. Part of the Ministry of Electric Power was transformed into the State Power Company of China (SPC) in 1997 to separate the enterprises from the administrative functions. Some other measures were also taken to improve the controllability and effectiveness of the power industry.

China's electric power industry was most radically reformed in 2002, when the State Council of People's Republic of China approved "The Scheme of Power Industry Reform." The SPC that owned about half the generation assets (the remaining assets belonged to IPPs and municipal bodies) and virtually all electric grids was divided into several companies, all of which remained state-owned. Five large generating companies were formed, whose power plants were distributed over many provinces, so that the share of each company in any local electricity market was no more than 20%. Besides, six regional grid companies were formed in the interconnected power systems of North, Northeast, Northwest, Central, and East China and also the southern and southwestern provinces.

The regional grid companies were arranged into two special companies:

- The South China Grid Company covered the south and southwest provinces.
- The State Power Grid Corporation covered the rest of the five regional grids.

The State Electricity Regulatory Commission was established in 2002 to ensure fair competition in the market. It was charged to develop market rules, to monitor and regulate market operation, and to maintain the market efficiency. The tariffs are fixed for each power plant individually for a long term and revised only at producers' request, bringing about an incentive and time for decreasing costs and gaining additional profit for producers. The tariffs for consumers are differentiated by category: residential, commercial, and industrial.

In addition to the functions of electric network development and maintenance, the aforementioned State Power Grid Corporation and the South China Grid Company perform the dispatching control and also the planning of generation capacity expansion. They determine the optimal time of commissioning, capacity, allocation, and type of new power plants, and announce a power plant construction auction. Investors who had won an auction would receive a guaranteed payment for capacity along with electricity sales revenues. Hence, part of the investment risk is shifted to the consumers and simultaneously the benefit from competition among producers is derived. Actually, the Corporation and the Company serve as the "Purchasing Agency" in the single–buyer market (though it is not indicated directly in [42, 81]).

Thus, China has arranged a market with regulated prices according to Model 2. Such a market allows China's power industry to develop at unprecedented rates—50–100 GW of new capacity was commissioned per year during 2004–2007. Construction of new power plants was financed from all possible sources: profits of the state generating companies, private investors, and municipal and probably state budgets. Nonetheless, there is still a power deficit in some provinces of China.

Despite sizable investments in power industry development and power deficit in some provinces, the electricity tariffs are maintained at a rather moderate level owing to regulation. By the data of [81] the tariffs for residents are around 5.4 ¢/kWh, for the commercial sector around 9.5 ¢/kWh, and for industrial consumers they range from 3.7 to 5.3 ¢/kWh plus payment for capacity. The latter consists of two parts: one is based on maximum demand (around $ 2.68/kW per month) and the other is based on the transformer capacity connected to the grid (around $ 1.83/kVA per month).

7.2.2 India

India is the second largest (after China) country in terms of population, with an intensively developing economy. The annual growth rates of electricity consumption in the last three decades have been about 7% [82], with permanent shortage of capacity and electricity reaching 10% and more. The country consists of many states (provinces) having a rather high autonomy and their own governments. The status of economic development (including the energy sector) in the states is highly diverse. Before 2002 the states owned vertically integrated power companies responsible for power supply to their territories. Besides, there are large-scale thermal, nuclear, and hydropower plants belonging to the Central Government that supply power for the corresponding states. The Unified Power System of the country is created using the back-to-back DC links between the four interconnected power systems, since they operate with different deviations of AC frequency.

Electric power industry reform in India was launched in 2003 with the adoption of the Electricity Act (EA 2003). This Act aimed to implement numerous measures and changes in functions of the governmental and regional authorities. Its main principle is to separate power generation from transmission in the states, where possible, and to create more favorable conditions for attracting private investments in the new power plants. At the same time the role of the Central Electricity Regulatory Commission that became independent of the Government was enhanced. The governmental and regional grid companies, with the possible inclusion of corresponding dispatching centers, were formed. The latter might be the independent departments under governmental control.

On the whole, the reforms in the power industry will be implemented according to the single-buyer market model, with regulated electricity tariffs.

The current status of reforms is different in different states (in some states there still are VICs), but in general the reform has substantially improved the situation in India's power industry. Introduction of the mechanism of payment to producers via the Availability-Based Tariff (ABT) was of particular significance [83]. Prior to 2003 the power deficit gave rise to frequency variation in power systems in absolutely inadmissible ranges (from 48 to 52 Hz), leading to repeated blackouts and load shedding. The ABT mechanism imposes an additional electricity payment for capacity increase by producers above the generation schedule, when the frequency falls below 50.5 Hz. The payment depends on frequency, increasing from zero at 50.5 Hz to 15 ¢/kWh at 49 Hz and below. The frequency and corresponding payment are measured every 15 min, and power plant personnel can follow and change power delivered in real time. Introduction of the ABT mechanism decreased frequency variation and emergency rates in power systems.

Though the reform in India was begun not long ago, the reform based on Model 2, with retained regulation of electricity tariffs, is believed to give positive results. It is absolutely impossible to introduce a competitive market (Models 3 and 4) with free prices because of capacity and electricity shortages existing there.

7.2.3 South Korea

The electric power industry reform in South Korea started in 1999, when the decision was made to restructure the monopoly state company KEPCO. This assumed a staged transition from Model 1 to Model 4.

In 2001, six PGCs were separated from the KEPCO, thus implementing the single-buyer model. Along with long-term bilateral contracts, a day-ahead market (DAM) was organized [84] where producers only competed. All PGCs received payment for electricity at marginal prices that were formed in the DAM, and payment for the capacity as well. The dispatching control was performed by the Power System Operator, which combined the functions of a market operator. Based on information available, the implementation of this regulated market resulted in a considerable benefit owing to tough competition that began among PGCs.

Yet, further restructuring of the KEPCO soon stalled. The privatization of one of the PGCs (KOSECO) and introduction of a competitive wholesale market (Model 3), which had been planned for 2003, failed. A considerable role here was played by the Tripartite Commission (Government, management, and trade unions) in 2003–2004. The Commission admitted that further division of the KEPCO (i.e., transition to Model 3) would not bring about any real benefits [85]. It should be noted that this was one of the rare cases in which the efficiency of electric power industry restructuring was analyzed and discussed so comprehensively and competently.

The electric power industry of South Korea continues to successfully develop at moderate electricity prices despite the fact that practically all the fuel for power plants is imported. Special importance is attached to the development of NPPs, CPPs on imported coal and liquefied natural gas, and pumped storage power plants (PSPPs). The necessity to construct capital-intensive power plants (NPPs, coal-fired CPPs, and PSPPs) in a competitive market could lead to an increase in prices up to the level required to pay back the investments. In a regulated market (Models 1 and 2) such construction is provided with an increase in consumers' tariff only by the averaged investment component (see Chap. 6).

Organization of competition among producers through the DAM (with an additional payment for capacity) at the single–buyer market should hardly be considered a successful decision (it was made when a further transit to Models 3 and 4 was supposed). Though in this case DAM differs from the spot markets under the competitive Models 3 and 4 (all PGCs sell electricity to the single buyer at the previously determined aggregate demand), its flaws mentioned in Sect. 5.2 persist. The producers have the possibility to increase the total price of electricity over their costs, and get excess profits by manipulating their DAM bids (and getting payment for the capacity). Supposedly, this was the case in South Korea—the wholesale prices gradually increased after 2001. Under fixed (regulated) retail prices the state company KEPCO (the "Purchasing Agency") appeared in a difficult financial position (unlike PGCs). There is information that the South Korea Government is discussing the issue of revising the wholesale market conception.

7.2.4 France

France belongs to a group of countries with poor energy resources. Therefore, since the 1960s and particularly after the world energy crisis of the 1970s, the country has been intensively developing nuclear energy. Currently, more than 80% of the total electricity consumed by the country and a considerable part of its export is supplied by the NPPs.

Almost the entire French electric power industry belongs to the monopoly state company Electricite de France (EDF). According to the recommendation of the European Council, the Law mandating electricity market liberalization was adopted in 2000, and the Transmission System Operator (TSO) and the Commission for Regulation in Energy (CRE) were formed. The Commission is a power system regulator that is assigned exclusive rights to organize and control market operations [86]. Its functions include the following:

- Approving the annual investment programs elaborated by the EDF and the electric network company RTE
- Approving new power plants, controlling power balance, and compensating imbalance
- Establishing electricity tariffs by taking into account power losses, taxes, capital value, and prices of connection to the power system
- Imposing sanctions on System Operators and consumers in the event that they violate the Commission's instructions

The Commission is obliged to annually submit a report to the French Parliament and the Supreme Council on Energy that describes its work, the results of power system operation, and the measures taken to connect consumers to the power system. The tariffs are corrected if needed, and when choosing a new power plant to be connected to the system the Commission adheres to the principle of the lowest price both for the System Operator and for the consumer.

Thus, the electricity market reform in France did not factually lead to complete deregulation and even streamlined power system operation under the state control.

The reliability of the power system operation in France is ensured by its regular control by the group of experts "Power System Reliability Audit Mission," set up by the System Operator. This group observes reliability and publishes annual reports about the reliability of power system operation, including a list of recommendations to improve its operation [87]. The rule of adequate reserve (N-k) is used as a criterion for reliable power system operation. In case of emergency, a System Operator must follow the established rules to restore the normal operation of the power system.

The retention of the state company EDF and its regulation provides successful development of the electric power industry and stable low electricity prices (as compared to the neighboring countries). France exports electricity to England, Spain, Italy, Germany, and other countries. Partial privatization of EDF is planned; however, the State will apparently retain a control package of shares.

7.2.5 Japan

There are about ten regulated vertically integrated private power companies in Japan. They supply power to their respective territories (prefectures). The power systems of the island of Hokkaido and the northern part of the island of Honshu operate at the frequency of 50 Hz, and the rest of the power systems operate at the frequency of 60 Hz. There are DC links between them with a limited transfer capability. Japan, like South Korea, has poor energy resources. It imports fuel for power plants and intensively develops nuclear power.

Reforms in the electric power industry of Japan were launched in 1995 when a special Act was adopted. The Act required monopolistic companies to buy electricity from independent producers. The latter, as a rule, are small power plants, including CGPPs, which are constructed to supply electricity and heat to industrial customers. In 2000, the Act was additionally extended. This resulted in the formation of retail electricity markets in some power systems where the prices for final consumers had been decreased due to IPPs, i.e., a positive effect was obtained.

In 2004, the market for the exchange of power among power systems was regularized, and a neutral organization, the Electric Power System Council of Japan, was formed. It contributed to developing rules and ensuring a fair and transparent market [84]. The measures were foreseen to eliminate electric lines congestion. Load plots and flows are calculated for the next day and next month, and are announced on a bulletin board similar to that used in the spot market and auction.

Priority is given to long-term contracts between companies. Consumers with a load of more than 50 MW at the voltage of 6 kV can choose a power supply company.

In spite of the very high electricity prices, Japan still has regulated VICs (Model 1) which provide normal expansion of power systems during the construction of expensive power plants, particularly NPPs. Introduction of a competitive market here would lead to an even greater increase in prices.

Box 20 Experience of the Countries with Regulated Electricity Markets

1. The countries that have retained the regulated monopolies or made a transition to the single-buyer market do not experience the drawbacks inherent in the competitive electricity markets. In particular, electricity producers do not have the possibility to gain excess profits: producer's surplus and monopoly profit.
2. For developing countries with high rates of electricity demand growth (China, India, etc.), the typical features are:
    a. Dominating state ownership in the electric power industry.
    b. The need to use private investments for expansion of the generation capacities. This can be provided by attracting the independent electricity producers by high interest rates or by a transition to the single-buyer market.
    c. The possibility to maintain moderate electricity prices at their regulation despite the great expenditures for EPS development and electricity shortage.
    d. The impossibility to make the transition to the competitive markets (price liberation) under the capacity shortage or necessity to construct capital-intensive power plants.
3. In the developed countries (France, Japan, South Korea):
    a. It may seem expedient to keep regulated monopolies (not to introduce the single-buyer market) at low rates of electricity demand growth.
    b. Transition to a competitive market is impossible if there is a need to construct NPPs, HPPs, environmentally friendly CPPs on coal, etc.

7.3 Experience of Implementing the Competitive Electricity Markets

After analyzing competitive electricity market operation over several years or even decades, experts started to divide them into "successful" and the rest ("unsuccessful" or with unclear results so far). In particular, the comprehensive work by the International Energy Agency [69] presents the analysis of four "successful" markets: Great Britain, Scandinavian countries, Australia, and the PJM market in the USA (the states of Pennsylvania, New Jersey, and Maryland). Among the remaining markets it is necessary to distinguish those that faced crises (obvious "failures") with restoration of electricity price regulation (as occurred in California).

The competitive markets in South America, Western Europe, and Australia will be discussed below. The PJM market is among the deregulated markets in the USA that were described in Sect. 7.1, along with the California crisis.

7.3.1 Brazil

The reform of the electric power industry started in 1997 with the privatization of state companies. In 1999, the competitive wholesale market was implemented (Model 3). The primary goal of the reform was to increase the state budget by privatizing power plants and to attract nongovernmental investments in the development of the electric power industry. Electricity prices were low due to the high share of HPPs (above 85%).

With the introduction of a competitive market the construction of new power plants stopped. Private investments in new HPPs did not pay back at low electricity prices, and it was quite risky to invest in the new thermal power plants, since during the high-water years they would be replaced by the HPPs and would not be paid back as well. During several years the market operated due to a drawdown of the HPP reservoirs with spot market prices fluctuating in the range of 0–9 ¢/kWh. Still electricity consumption increased and in 2001, with the HPP reservoirs depletion and low water in the rivers of southeastern Brazil, the country faced an electricity shortage which caused spot price to bounce to 50 ¢/kWh [8, 88].

The Government asked all the consumers (including residents) to curtail electricity consumption by 20% which was done quite promptly. This, along with subsequent high-water years, made it possible to overcome the crisis.

Simultaneously the market structure was changed:

- The DAM, which had been used earlier for almost the whole electricity trade, was abolished.
- The regulated sector of the wholesale market was formed with the trade exclusively by the long-term bilateral contracts between the PGCs and DSCs.
- The free trade sector was retained, but also with the long-term bilateral contracts only (at unregulated prices); the participation in the sector is permitted to IPPs, PGCs (over and above the contracts signed with DSCs in the regulated sector), and the so-called free consumers and traders, but the DSCs with regulated retail prices cannot take part in the sector.
- The balancing market is retained with spot prices calculated by the special models (i.e., the market prices are not formed as equilibrium ones).

All participants of both market sectors (sellers and buyers) are obliged to provide 100% of electricity volumes to be produced or consumed under long-term contracts both for a short- and long-term perspective. Various auctions are held among operating and new producers which create competition among all electricity producers, including new ones. The auctions among the *operating* producers are held at least 1 year in advance, and the contracts with winner producers are signed for a term of 3–15 years. The auctions within the *new* producers should be held well in advance of 3–5 years, and the term of contracts is 15 years for thermal power plants and 30 years for HPPs. Electricity is purchased from the producers at the long-term prices claimed by them during the auctions. Volumes of electricity (and power) delivered by producers to the regulated market sector are distributed afterwards between the separate DSCs. Consumers' tariffs are regulated (averaged) inside each DSC.

About 70% of electricity is now sold through the regulated wholesale market sector (and 30% through the free sector). Spot prices in the balancing market are now around 8 ¢/kWh (operating producers have lower prices stated in the contracts, especially for HPPs, but new producers could have higher prices).

The experience of Brazil in overcoming the deficit and subsequent reforming of its electric power industry is worth studying. In fact, they have now a kind of single-buyer model (in the regulated sector).

7.3.2 *Argentina*

The reform of the electric power industry started in 1993 [8, 88] as part of a wider reform of the national economy. The state's monopoly power companies were split, partly privatized or given on concession. Initially, the reform brought a considerable positive effect. The wholesale electricity prices decreased from about 5 ¢/kWh in 1992 to less than 2.5 ¢/kWh in 1997, despite the 5.7% increase in average annual electricity consumption. Power systems expanded through the construction of the gas turbines (before 1997) and then CCPPs, while the construction of HPPs had been stopped. Electricity market in Argentina was considered to be a successful one, even a paragon.

The situation changed dramatically in the end of 2001 due to severe political and economic crisis in the country which, in particular, caused a threefold devaluation of the national currency (peso) against the US dollar. Prices and tariffs in most contracts with domestic and foreign investors were indicated in US dollars, which caused problems with the payback of investments and could lead to a manifold increase in the internal electricity prices. The Government had to introduce the regulation to overcome the crisis, thus eliminating the competitive market.

Simultaneously, private investments in the construction of new power plants had stopped, and in 2004 a special state company ENARSA was founded. It is responsible for the development of the energy sector in the country, including the electric power industry. This company has played an increasingly important role competing with other companies, including private ones.

7.3.3 Chile

Chile was the first country to begin the reform of the electric power industry (in 1982) and create a competitive wholesale market [8, 88]. For more than 10 years the reform succeeded, particularly, in enhancing the electricity production efficiency (decrease in prices) and attraction of private investments. CCPPs developed most intensively. They operated on cheap natural gas imported from Argentina, meeting the rapidly growing electricity demand. At the same time in the 1990s several crises occurred caused by a low-water period at the HPPs. They were accompanied by the spot market price increase and the emergency disconnection of consumers. The most severe crisis of this kind occurred in 1998–1999 when the HPP water reservoirs were completely exhausted.

Dramatic changes occurred in 2004 when the Government of Argentina decided at first to curtail and afterwards to stop gas export to Chile [8] due to the internal economic problems. The construction of CCPPs on cheap gas from Argentina turned out to be impossible, and even operating power plants suffered gas shortages. The only fuel resources at hand were coal and fuel oil, which fortunately could also be used at the existed CCPPs. Alternative energy resources (imported liquefied natural gas, new coal-fired power plants, HPPs at remote Patagonia) required a lot of time to be developed. With growing electricity demand, the country faced power shortages and spot prices bounced up to 30 ¢/kWh.

The Government had to change the concept of electric power industry reform. Based on the available data, Chile intends to follow the example of Brazil, i.e., to introduce the state regulation of the market for DSCs with transition to the long-term contracts with operating and new power producers to be signed on an auction basis, and retain the competition sector for "free" consumers.

7.3.4 Great Britain

Similar to the majority of other countries in Western Europe, the reform of the electric power industry in Great Britain started under very favorable conditions: large reserves of capacities at slow rates of electricity consumption growth, the possibility to widely utilize cheap natural gas in combined-cycle installations, fairly developed electric networks, etc. The reform was accompanied by the privatization of the electric industry, which was previously entirely state-owned. Power plants were partly privatized in 1990–1991 with the forming of three large PGCs, and till 1996 all electrical facilities including NPPs, HPPs, electrical networks, and the sphere of sale were privatized. It should be mentioned that the original 12 regional electricity distribution and sales companies were gradually transformed during the reform into 6 large sales companies by 2003 [69]. Some of them have their own generation, and the British Gas Company sells both gas and electricity. Substantial amount of the British electricity industry companies' stocks were bought by foreign companies (from the USA, Germany, France, etc.)

The British competitive market started to function in 1990. It was organized according to the "classical" at that time conception (similar to that applied earlier in Chile, and later in Argentina, USA, Brazil, and some other countries). All trade was performed through a spot day-ahead market at equilibrium prices. The producers got additional payment for the available (used in hourly balances) capacity, which was calculated by a certain formula depending on the "Loss of Load Probability" and on "Value of Lost Load."

After creation of the competitive market the electricity production efficiency was enhanced, and electricity prices decreased in the beginning of the 1990s. Yet, this was caused not only by the competition but by many other factors as well, for example, replacement of coal-fired power plants by the combined-cycle ones, decrease in the natural gas prices, preliminary compensation for stranded costs, etc. [89]. These factors would decrease the prices under the regulated monopolies as well. As to the benefit obtained during the first years of deregulation it went mainly to power producers. In [95] the author notes that the prices exceeded the costs so much that within just one year the shareholders of the PGC National Power got dividends exceeding the primary value of the company when privatized. According to the data presented in [9], as a result of the reform, the electricity producers gained 9.7 £ billion, the Government gained 1.2 £ billion, whereas the consumers lost 1.3 £ billion. Hence, the reform brought a net benefit but not to the customers who lost money.

Transition to the competitive market coincided with the large-scale development of the natural gas resources in the North Sea shelf and creation of efficient CCPPs. As was mentioned in Sect. 6.4, to recoup investments in new CCPPs on cheap (at that time) natural gas the wholesale prices had to be approximately 3.8 ¢/kWh. The actual wholesale British market prices (including payments for capacity) were higher, and this stimulated intensive development of such power plants. Coal-fired power plants appeared to be incompetitive, especially after the Government stopped to support its own coal industry. The CCPP construction "boom" was accompanied by the closure of obsolete coal-fired power plants. At the same time most closed plants were not dismounted but retained in "cold" reserve for a case of power deficit. On the whole, taking into consideration the slow increase in demand, this practically alleviated the problems of the generation capacity development (the annual load peak at the market area increased by some 4.5 GW in 1991–2004, or less than by 10%).

Meanwhile, the drawbacks of the first market conception started to appear. Wholesale prices diminished slower than generation costs, and in some years they even increased. The use of market power by producers, their manipulations of price bids in the spot market with receiving payments for capacity became evident. In 1997–1998 the Electric Industry Regulation Authority had analyzed the market functioning, presented their critical remarks, and made recommendations for a crucial change of the market conception. After this the New Electricity Trade Arrangement (NETA) was developed, and implemented in March 2001. The main features and differences of NETA from the original market conception are as follows [69]:

- Transfer of all electricity trade to bilateral long-term contracts (at the free contracted prices) concluded for a period of several years and liquidation of the spot day-ahead market.
- Implementation of a balancing (hour-ahead) market with the mandatory bids of all the market players. Inside the national network company a daughter company ELEXON was formed which controls the function and accounting at the balancing market.
- Absence of any special mechanisms for generation capacity development, including payment for capacity.

Unlike other spot markets, the NETA conception implied in the balancing market not the marginal but the so-called discrimination price formation.[4] This means the market players sell and buy electricity at prices shown in their bids, not at the equilibrium prices.

A task had been put up to create an exchange forward market of standardized long-term contracts. This market would be *a true competitive market*, which is the only one to be formed theoretically in the electrical industry. The electricity trade would be provided in this market at the prices reflecting the *total* producers' costs (including the fixed ones), not the variable ones (hourly) as it takes place in the spot markets in real time. At the same time it would send the required "price signals." Still some obstacles had been encountered at its way, that's why it has not been formed till now. Bilateral long-term contracts are signed beyond exchange areas, by means of separate agreements between producers and buyers (consumers). The prices in these contracts are confidential, and no "price signals" appear.

Scotland was included in the British market in 2005, and the NETA mechanism was transformed into the British Electricity Transportation and Trade Arrangement (BETTA). All the main structures and rules of the NETA mechanism were retained in BETTA.

The British electricity market as a whole can hardly be considered "successful" (especially for consumers). It is thought to be such in [69] maybe due to the fact that there were no such crisis phenomena in Great Britain as in California, Brazil, Argentina, and Chile. However, the original market conception was radically changed, and the NETA (BETTA) conception has not been completely realized yet. One can expect further changes of the reform conception, since there are new tendencies in the electric power industry of the country. In particular, generation companies merge with sales companies (and expand), i.e., their vertical integration and monopolization occur "bypassing" the competitive wholesale market [2]. Besides, sooner or later the problems of generation capacity development will arise and call for solution (it is impossible to transfer the entire electric power industry to natural gas).

---

[4] Such price formation should be rather called "fair" with regard to buyers, as the marginal prices are formed according to the most expensive accepted sellers' bid, thus creating profit for the other sellers and increasing buyers' expenses.

7.3.5 Scandinavian Countries

The market of Scandinavian countries (NORDEL) is an interstate one. It allows the countries to gain the benefits from interconnection of the countries with different structures of generation capacities. It is particularly efficient for Norway, which mainly operates HPPs, owing to electricity export in high-water years and seasons and import during low-water periods. However, considering Norway separately, its internal competitive market could hardly bring the benefit due to low electricity prices before the reform, and a peculiar structure of capacities, with 90% of relatively small HPPs.

The NORDEL market started to function in 1996 in two countries—Norway and Sweden [69]. In 1998 Finland joined it, in 1999 West Denmark, and in 2000 East Denmark. The market structure is slightly different in these countries, but the main common sector is the DAM (Elspot) with the marginal zonal price formation. Finland has allocated five zones, the other countries are represented by one zone each. There is also a balancing market (Elbas), with the hour-ahead bids for participation in it. Substantial part of the trade is based on bilateral contracts. The derivative (options, futures) financial markets are well developed. The firms and companies that are not connected to electrical networks of the market can take part in those derivatives. About half the electricity trade is carried out through the DAM (Elspot), the other part through bilateral contracts. However, the latter are concluded by the producers mostly with the sales companies owned by them.

Many of the Scandinavian electricity companies are state-owned or municipally owned, and in Norway almost all companies are so. The Danish Government supports the RES installations construction, especially the wind farms, which share exceeds 30%. They require full backup of their capacity; thus, the Danish power system reserve increased to about 100%.

Payment for capacity and any mechanisms for expansion of generation capacity are not provided in the NORDEL market. The prices formed during almost all the years of reform were lower than those necessary to pay back investments in the new power plants, even in the CCPPs.

The NORDEL market (as that of Great Britain) is considered to be successful because it has not yet suffered any crisis. In particular, it managed to "survive" the drought that occurred in Scandinavia in 2002–2003. At the same time, there are facts and trends that can be considered flaws of this market:

- With the introduction of competition in Sweden and Norway, electricity prices started to increase (instead of decline) much faster than the general index of consumer prices [4, 9, 60].
- The investment problems arose, and the construction of new power plants and networks almost stopped [4, 9, 69]. As a result, capacity reserves decreased (apart from Denmark) which caused bottlenecks and congestion in some branches of the network. The only new NPP Oekiluotot (Finland) is constructed with the

money of large consumers and for their supply (in fact, it will not be involved in the electricity market).

- On September 23, 2003, a long blackout occurred in Sweden and Denmark, where 4 million people lost electricity for half a day. By the way, in the same year on August 28, a short-term blackout occurred in London.
- Electricity export under a competitive market is not profitable for the consumers of the country-exporter, as was shown in Sect. 6.5. This caused damage to the Norway and Sweden electricity consumers where the prices before the reform were the lowest in Western Europe.

Thus, the electricity market in Scandinavian countries cannot be considered as fully successful, at least for consumers. Like the other competitive markets it is getting more sophisticated (implication of derivatives etc.). One would expect that the major troubles are likely to emerge after the available capacity reserves are exhausted and the need arises to restart the construction of new power plants.

7.3.6 Other Western European Countries

It is rather difficult to review electricity markets of all countries, and therefore we will dwell on some facts only.

In Germany, shortly after the implementation of a competitive market an unjustified increase in electricity price began [53, 60]. A particularly difficult situation in the electricity market exists in Italy which has suffered severe electricity deficit covered by imports from neighboring countries. Italy has the highest electricity prices, and on September 28, 2003, a severe blackout occurred that affected the whole country: 57 million people had no electricity supply for 12–24 h.

On the whole, the efforts and directives made by the European Union to create a single European electricity market have been implemented with great difficulties and delays and not in all the countries [2]. This can be explained from our viewpoint by two factors arising during the transition to the competitive market, which were discussed in Sect. 6.5: electricity export ceases to be mutually profitable and challenges emerge to finance and to ground the effectiveness of interstate transmission lines.

The requirements of the Kyoto Protocol, CO_2 emission quota trade, and a tendency to develop RES have started to play an increasingly more important role in the last years. Even greater difficulties can be expected in several years, when the need arises to construct new capital-intensive power plants due to electricity demand rise (though not very high), and decommissioning obsolete power plants.

7.3.7 Australia

The National Australian market (competitive) started to function in December 1998 in the states of New South Wales (including the Capital of Australia area), Victoria,

Queensland, and South Australia [69, 91]. Later, in 2006, the island of Tasmania was affiliated by a high-voltage submarine cable.

The vertically integrated state companies had been in advance restructured with the creation of the independent PGCs, the national network company (NEMMCO), which simultaneously realized the System Operator and Trading System Administrator functions, etc.

The national market is based on the spot day-ahead market with the whole electricity trade provided. Long-term bilateral contracts and payment for capacity are not used. There are the ancillary services markets for the support of frequency, reserves, voltage, etc. A market (exchange) for derivatives was formed with the trade by the Contracts for Difference (CFD), assured to soften large volatility of prices in the spot market.

The spot market prices are formed by zones (nodes) coinciding mostly with the state areas due to limited interstate transmission lines capability. The Snow Mountains zone is allocated in addition. This zone is situated in two states (New South Wales and Victoria). Electricity prices can be sufficiently different in different zones.

The spot prices in the state of South Australia immediately bounced after the competitive market had been introduced (since January 1999). This state suffered electricity deficit, and imported it partly from the neighboring state of Victoria. Monthly average prices made up 95, 105, and 135 AU $/MWh in November 1999, February 2000, and February 2001 respectively at a “normal” price of about 30 AU $/MWh. In 1999–2000, the average price was 61 AU $/MWh. High prices fostered construction of 300 MW simple-cycle gas plants and of 800 MW combined-cycle gas plants, which augmented the installed capacities of the state by 30%.[5] After this, since June 2001 the prices had been diminished to a normal level. However, a two-year price bounce, of course, caused losses to consumers, and brought excess profits to operating producers.

A more drastic crisis burst out in 2000–2001 in the state of Victoria, which had more than 30% of power reserve and exported electricity to South Australia. Absence of new power plants and continued demand increase during the hot summer of 2000 (January–February) led to the electricity supply failures and bounce of the spot prices. Some “fan-shaped” consumer trippings occurred. The State Government introduced spot price cap, and afterwards the peak hour electricity consumption limits. The crisis was overcome in March–April 2001, after the new 650 MW capacities had been commissioned.

After the crisis phenomena in the states of South Australia and Victoria which coincided in time with the NETA implementing in Great Britain, the national market conception was revised and partly corrected. In particular, a new national entity, the Australian Electricity Market Commission, was introduced in 2005. It is responsible for the working out of norms and rules of market functioning and developing. In parallel with this, the Australian Energy Regulating Administration was formed

---

[5] As was already mentioned in Sect. 6.4 the price necessary for the recoupment of the new CCPP in Australia comprised around 40 AU $/MWh.

to effectively regulate the wholesale market and transmission networks in the electricity and gas markets [69]. However, the crucial change in the electricity reform conception similar to British NETA and BETTA has not taken place so far. The spot day-ahead market and other related markets are retained. No mechanism for expansion of generation capacities has been created.

The national electricity market in Australia might be evaluated as "successful" if the aforementioned crises in the states of Southern Australia and Victoria are not considered. However, the generation is developed only at the expense of GTPs and CCPPs on natural gas, and the market improvements continue.

Box 21 Experience of the Countries with Competitive Electricity Markets

1. After transition to a competitive market practically all countries stopped construction of HPPs, NPPs, coal-fired CPPs, and intersystem and interstate transmission lines.
2. Energy crises occurred in Brazil and Argentina where state regulation was restored. Crisis phenomena related to capacity shortage, spot prices bounce, and consumers curtailments took place in Australia (the states of Victoria and South Australia), in Chile, and in other countries.
3. Despite the favorable "starting" conditions (large capacity reserves at low rates of electricity demand growth, well-developed electric networks, etc.), after transition to a competitive market in the countries of Western Europe, including Great Britain, the drawbacks of the competitive market started to appear:

   a. Increase in electricity prices in Finland, Sweden, Germany, and other countries (outpacing a general index of consumer prices).
   b. Deficient investment in the new power plants and electric networks.
   c. Blackouts in Sweden, Denmark, Italy, and England in 2003.
   d. The main effect of market deregulation is gained by electricity producers.

4. Great Britain eliminated the spot day-ahead market in 2001, and started to trade under the long-term bilateral contracts, thus retaining the balancing market only. Under the balancing market, however, they use "discriminative" pricing and do not use the marginal one, i.e., payments are effected at the prices that were indicated in the bids of market participants.
5. Efforts and directives of the European Union aimed to create a single European electricity market are implemented with great delays, and not in all countries. It is mainly explained by the fact that electricity export in the competitive market becomes unprofitable for the consumers in the exporting country and producers in the importing country.

6. Competitive markets grow increasingly complicated, similar to the situation in the USA.
7. The crises caused by the shortage of generation capacities (similar to Brazil and Chile) may be expected to occur in other countries as well, after the possibilities of using cheap natural gas are depleted.

Chapter 8
Power Industry Reforms in Russia

This chapter deals with conditions, goals, course, and potential results of electric power industry reforms in Russia based on the foregoing information. To gain a full picture, the initial restructuring at the country's transition from the planned economy to the market one in the 1990s (Sect. 8.1), the next reform related to the transition to the competitive market (Sect. 8.2), and a forecast of its consequences that can be made for the coming decade (Sect. 8.3) are considered.

8.1 The Reform of the 1990s

Disintegration of the USSR and transition to the market economy in the country gave rise to the "total" privatization along with the creation of joint stock companies. In compliance with the Decree of the President of the Russian Federation of 01.07.92 No. 721 "On organizational measures of transforming state enterprises and voluntary associations of state enterprises into stock companies," such work should be carried out during 4 months and completed before November 1, 1992. For the power industry, it meant a complete breakdown of the Unified and Regional Power Systems into a multitude of self-governing joint stock companies (power plants, network enterprises, etc.).

Efforts of the administration and experts of the Committee on electric power industry of the Ministry of Fuel and Energy of the RF contributed to the regulation of the process of creating joint stock companies [92]. Another Decree of the President of the RF (No. 923 of August 15, 1992) "On the organization of management of the electric-energy complex of the Russian Federation under privatization" was prepared. According to this Decree:

- The Russian joint stock company of energy and electrification (RAO "EES Rossii") is established.
- Managerial bodies of regional power systems are transformed into subsidiary companies (AO-Energos).

L. S. Belyaev, *Electricity Market Reforms*,

DOI 10.1007/978-1-4419-5612-5_8,

- Condensing power plants with a capacity of 1,000 MW and higher and hydro power plants with a capacity above 300 MW are removed from regional power systems and transformed into subsidiary companies (AO-Elektrostantsiyas).
- Central Dispatching Board, Regional Dispatching Centers, design and research institutes, educational institutions, and construction and erection organizations of the industry are transformed into joint stock companies and contribute shares fully or partially to the capital stock of RAO EES Rossii.
- All cogeneration power plants remained in their AO-Energos.

As a result of the implementation of this Decree, 74 AO-Energos and 36 AO-Elektrostantsiyas were created. Two AO-Energos were independent of RAO EES Rossii: "Irkutskenergo" and the nonprivatized State Unitary Enterprise "Tatenergo." From 14 to 100% of shares of the rest of 72 AO-Energos belonged to RAO EES Rossii. Some AO-Elektrostantsiyas were leased out to the corresponding AO-Energos.

On the whole, the described scheme of transforming power industry into joint stock companies aimed to organize a federal single-buyer market and create regulated monopolies at the regional level (Fig. 8.1). It was supposed to organize the federal wholesale market of electricity and capacity (FOREM), to which AO-Elektrostantsiyas, NPPs, and surplus AO-Energos supply electricity. RAO EES Rossii, as the FOREM organizer, also acts as a Purchasing Agency. Tariffs for the electricity supplied and purchased in FOREM are regulated by the Federal Energy Commission (FEC). Tariffs for consumers supplied by AO-Energos are fixed by the Regional Energy Commissions (REC).

Transition of the power industry from a centralized planning to a market environment was performed wisely enough, though in the shortest possible time. The administrative-economic integrity of the UPS of Russia and regional EPSs was retained, and the regulated electricity markets were organized. Certainly, it would be reasonable that the power industry be state-owned, as in France, Norway, China, and many other countries. However, for Russia it was impossible (except for nuclear energy) because of the conditions prevalent there at that time. Since the

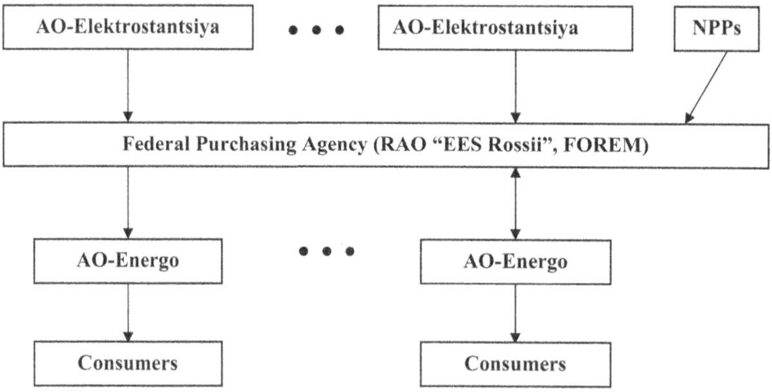

Fig. 8.1 A two-level structure of regulated electricity markets created in Russia in the 1990s

controlling block of shares of RAO EES Rossii remained state-owned, the Government had sufficient levers for industry management, including the state electricity tariff regulation. It should be noted that privatization of the power industry (as well as other industries) was carried out "free of charge," i.e., new owners of enterprises (shareholders) were not to pay or refund in any form capital investments done earlier (this aspect has already been mentioned in Sect. 6.4).

Further course of Russia's power industry development and operation in the 1990s was under the influence of some factors and circumstances:

- The general economic crisis in the country and electricity consumption decline
- The necessity to create a system of state regulation of energy companies (FEC and REC) that appeared as a new form of activity
- Insufficiently complete implementation of the single-buyer market model in FOREM
- Change in the top management of RAO EES Rossii in 1998

The economic crisis caused very difficult conditions in the power industry. Together, inflation and nonpayments violated financial and economic activity of energy companies. Devaluation of the fixed assets led to the understating of a depreciation component of tariffs. Therewith it had to be spent not for equipment renewal, but for other daily needs. Arrears of payment for fuel supply and personnel wages arose. Energy companies gained virtually no profits, the dividends for shareholders were not paid, and the employees (of energy companies) having shares sold them for nothing. Currently these shares are the property of different companies and banks, including foreign ones.

All industry indicators gradually deteriorated: fuel consumption per 1 kWh of generated electricity, network losses, number of personnel, capital investments, etc. Modernization and replacement of obsolete equipment at power plants and networks were carried out to a much lesser extent than necessary. Commissioning of new capacities in 1992–2000 made up nearly 10 GW, i.e., five to eight times less than that in the 1960s–1980s. All these capacities were used to compensate for the removed obsolete power plants, and as a result, the total installed capacity of power plants in Russia in the 1990s virtually did not change. Network construction also decreased sharply.

Decline in capital construction led to the degradation of construction basis, energy machine building, and design and engineering enterprises. The situation was somewhat mitigated by a general decrease in the electricity consumption (by 23.6% by 1998). The capacity reserves emerged, which created the illusion that everything was fine. However the volume of obsolete equipment continued to grow, thus worsening the situation for the future.

A general economic crisis extremely complicated the activity of *state regulatory bodies* that had to be started anew. It was very difficult to regulate tariffs of energy companies under high inflation, nonpayments, and accounts receivable and payable. This was accompanied by political and social factors that made the regulatory bodies restrain the growing tariffs, particularly tariffs for consumers at a regional level.

Electricity tariffs had to be revised very often (sometimes several times a year). Therefore, it was impossible to create incentives for regulated companies to decrease costs and gain additional (economic) profit. As was shown in Sect. 4.2, this was possible only if tariffs for producers were fixed for quite a long period of time (several years). In China, for example, in the absence of (or at very low) inflation, the tariffs are revised only on the initiative of the producers.

Inefficient changes in the legislation in 1997 should also be noted. These concerned withdrawal of an investment component from electricity tariffs for AO-Energos and transition to financing the development of regional power systems from profits of power companies. This disturbed the normal process of self-financing in regional systems and in fact excluded the possibilities to influence this process and control it by REC. Luckily, the investment component was preserved in the subscription fee of RAO EES Rossii.

Despite all these difficulties, the system and methodology of state regulation was gradually improved though not all proposals of the Federal Energy Commission were accepted.

Insufficiently complete implementation of the Single-buyer market model in FOREM first of all lied in the fact that AO-Elektrostantsiyas and many *surplus* AO-Energos that supplied electricity to FOREM were not *independent* producers. They belonged to RAO EES Rossii as daughter companies. Regulated FOREM in fact served as a mechanism of averaging the wholesale electricity prices. RAO EES Rossii that owned AO-Elektrostantsiyas and almost all AO-Energos was the monopolist in FOREM. There were many known cases where access to FOREM of nuclear power plants and AO-Energos that did not belong to RAO EES Rossii was hampered. There was no real competition among producers in FOREM as it can be in the single-buyer market.

Second, in the 1990s for the foregoing reasons the tariffs of electricity supplied to FOREM by AO-Elektrostantsiyas had to be often revised. They were based on the actual costs which eliminated incentives for producers to decrease them. However if the tariffs were established for a long period and producers were independent indeed, the effect of competition among producers for entry to the market and their quest for maximum profit (already discussed in Sect. 4.2) could be realized. This became possible only in 2000–2002, and instead of transition to the competitive market the efforts had to be made toward improvement of the state regulation and complete implementation of the single-buyer model. In 1997 an attempt was made in this direction. The Decree of the RF President No. 426 of April 28, 1997, about restructuring of natural monopolies was issued. It suggested, in particular, the creation of independent generation companies. However, this Decree was not fulfilled.

It should also be noted that a two-level system of regulated markets that was created in the 1990s, in principle, made it possible to attract private (external) investors for construction of new power plants (along with the use of investment components in tariffs). RAO EES Rossii and AO-Energos could conclude long-term contracts with independent private investors. The contracts stipulated increased prices of electricity bought from investors, which provided recoupment of investments at a mutually acceptable interest on capital. This was practiced in China as far back as in the 1980s. However unstable economic position of Russia created a very high risk for investors. Besides, decline in electricity demand and, as a result, emergence of generation capacity reserves made commissioning of new capacities temporarily unnecessary. The major problem by the end of the 1990s was modernization and upgrading of equipment at operating power plants as well as restoration of general operating efficiency of UES and regional power systems.

Change in the top management of RAO EES Rossii in 1998 was, to a great extent, a "subjective factor," which, however, had a rather essential impact on the further development of the Russian power industry. The change occurred in the critical period for the country (including 1998 default) when it became clear that power industry had already been in crisis. It seemed that the new administration of RAO had to take measures to improve the situation, but this did not happen.

RAO EES Rossii was headed by laymen, i.e., managers (economists, lawyers, etc.) who had nothing to do with energy. During several years highly professional energy experts in the daughter AO-Energos and AO-Elektrostantsiyas were also replaced by managers. The major concern of the energy companies' administration became business. This manifested itself to the greatest extent in the first years of the twenty-first century (discussed in the next section).

Of special importance was the fact that the new administration of RAO EES Rossii saw the way to electric power industry recovery from crisis in its further restructuring instead of implementing concrete and fast measures on improvement of management and technical upgrading of electric power industry. Thus, the recovery was postponed for 5–10 years more.

It can be supposed that if the top management of RAO EES Rossii still had the professional energy experts who managed to maintain integrity, operability, and reliability of UPS in the most difficult years (i.e., 1992–1998), Russia's power industry would have developed in another way. The problems of aging equipment, degrading construction basis and machine building would not have been exacerbated, unjustified costs would not have been borne, and new problems would not have arisen.

Box 22 The Reform of Russian Electric Power Industry in the 1990s

1. With transition of the country to the market economy, the privatization (creation of joint stock companies) was carried out in the power industry. Owing to the efforts of energy experts, the economic integrity of the Unified Electric Power System (UPS) of Russia and regional power systems was preserved. A two-level structure of regulated markets was created: the single-buyer market at the federal level and regulated vertically integrated companies at the regional level.
2. The general economic crisis created a very difficult situation in the industry. Inflation, nonpayments, depreciation of assets, etc., interfered with the financial and economic activity of energy companies. All the indices of the industry gradually deteriorated and reached a critical level. The decrease in electricity consumption and high organizational and technical level of UPS

that had been achieved by early 1990s somewhat mitigated the situation. However, the problem of equipment aging at power plants and networks grew increasingly urgent.

3. The system of electricity tariff state regulation had to be created from scratch under difficult conditions of economic crisis. It was extremely difficult to regulate tariffs under high inflation, nonpayments, accounts payable and receivable by power companies. The tariffs were often revised and therefore the companies had no time and incentives to decrease costs. The situation was additionally complicated by political and social factors that led to the urge toward reduction of tariffs for electricity. At the same time the state regulation of tariffs was gradually improved.
4. The single-buyer market model was implemented in the federal wholesale power market (FOREM) incompletely, as far as many AO-Electrostantsyas and the majority of AO-Energos participating in the wholesale market were not independent. They were affiliated companies of RAO EES Rossii that in reality was the monopolist in FOREM.
5. Change of the top management in RAO EES Rossii in 1998 unfavorably influenced the ways of overcoming the crisis in electric power industry. The energy experts were replaced by managers (economists, lawyers, etc.) whose main concern became business. Instead of concrete measures aimed at enhancing the effectiveness and re-equipment of the industry, the new administration of RAO began elaborating suggestions on its further restructuring, postponing the measures on overcoming the crisis for 5–10 years more.

8.2 Further Restructuring with Transition to the Competitive Market

By 2000–2002 the financial situation of RAO EES Rossii and AO-Energos stabilized owing to a general improvement of the situation in the monetary system of the country and elimination of debts of consumers (particularly, budget organizations). Fixed assets were re-estimated, which increased a depreciation component of tariffs. A *large-scale* renewing and updating of energy facilities could be started, constructions commenced earlier could be completed, etc. With a two-level system of regulated markets, this could be well done at the expense of depreciation and investment components of tariff.

By that time, several large works on perspective development of Russia's electric power industry till 2010–2020 were performed. These are the work supervised by the Krzhizhanovsky Energy Institute (KEI) [49], Energy Strategy of Russia till 2020 (ESR) [93], and the studies conducted at the Institute of Economic Forecasting

Table 8.1 Capital component of tariff, ¢/kWh

Work	2005	2010	2015
KEI [49]	0.4	0.86	1.27
ESR [93]	0.31–0.47	0.61–0.73	0.77–1.23
IEF [94]	0.47–0.52	0.82–0.83	–
Average estimate[a]	0.44	0.77	1.09

[a] Arithmetic average by column

of RAS (IEF) [94]. The works presented the forecasts of electricity consumption, variants for expansion of generation capacities (with account taken of updating and dismantling of operating power plants) and electric networks, demand for capital investments, etc. The authors of [19] used these works to make the generalized estimation of a *capital component* of tariffs, which is necessary to modernize the existing power plants and construct new ones in the last years of 5-year periods. These estimates are presented in Table 8.1.

As is seen for the electric power industry to recover from crisis at that time, it was necessary to increase electricity tariff by less than 0.5 ¢/kWh in 2005, by 0.7–0.8 ¢/kWh in 2010 and by nearly 1 ¢/kWh in 2015. The capital component rises due to an increasing volume of obsolete equipment to be replaced. The large, if not the larger, part of this capital component is covered by the depreciation charges included in the tariff in any case. This would have been conducive to updating and construction of power plants and networks, operation of energy machine-building factories, work of construction, erection and design companies, etc.

Instead, in December 2000 the new administration of RAO EES Rossii submitted the concept of restructuring RAO EES Rossii to the RF Government for approval. The necessity of restructuring was substantiated by the critical state of the Russian power industry and it was suggested as a remedy for recovery from the crisis. The document proposed transition to the *competitive* market (Model 4) in Russia's power industry. Some organizations [95, 96] were involved in the development of the concept.

The concept of RAO was thoroughly discussed and criticized. By the resolution of January 7, 2001, of the RF President a Working Group on Power Industry Restructuring was created at the Presidium of the RF State Council to consider the concept. On February 23, 2001, Parliamentary hearings "On the situation in electric power industry and restructuring RAO EES Rossii" were held in the RF State Duma. In February 2001 a joint meeting of the three departments of the Russian Academy of Sciences was held (Department of Physical and Technical Problems in Energy, Department of Geology, Geophysics, Geochemistry and Mining Sciences, and Department of Economics). The participants of the meeting admitted that the concept of RAO EES Rossii could not be taken as the basis for the state policy on restructuring Russia's power industry.

The author of [58] described in detail the discussion of the concept. About ten alternative concepts were proposed including that put forward by the Working Group at the Presidium of State Council. However, the RF Government ignored the proposals of experts by issuing Decree No. 526 [1] of July 11, 2001, and thus approving

"The Main Directions of Electric Power Industry Reform," which virtually completely corresponded to the concept of restructuring RAO EES Rossii. This Decree initiated the transition of the electric power industry to the competitive market.

At the end of 2001, the RF Government introduced a draft law on the electric power industry to the RF State Duma. It was also based on the concept of RAO restructuring. The draft law was actively discussed and parliamentary hearings were held. As a result certain amendments were made by the State Duma in the first and second readings. They were mainly related to strengthening the State and Government role in implementing the reforms, controlling their results, and ensuring continuous power supply. The final version of the Law "On the electric power industry" was adopted by the State Duma on February 21, 2003, and signed by the RF President on March 26, 2003 [39]. Simultaneously the Law "On the specifics of electric power industry functioning during the transition period" [97] and some related laws were also adopted.

The laws [39, 97] emphasize the *transition period*, i.e., the time period before the wholesale market rules come into force and price regulation in the wholesale market is terminated. The transition period was planned to end no earlier than July 1, 2005. Later, this date was postponed many times, and according to the last resolutions of the Government, it was prolonged to 2010.

Let us first consider the *goals of power industry restructuring and possibilities for their achievement* in the context of materials presented in Chaps. 4–7. The Law "On the electric power industry" [39] does not state the goals of the reforms clearly, but there are "general principles" (Article 6) and part of these can be considered as goals:

- To provide energy security of the Russian Federation
- To ensure continuous and reliable functioning of electric power industry
- To form a stable system to meet the demand for electricity, provided the appropriate quality and the electricity cost minimization is ensured

The Decree of the RF Government No. 526 [1] (to be more precise the approved "The Main Directions of the Electric Power Industry Reform") reads:

> The goals of Russia's power industry reform are to provide sustainable functioning and development of the economy and social sphere, enhance the efficiency of electricity production and consumption and ensure reliable and continuous power supply to consumers.

Additionally, according to Decree No. 526, it follows that one of the reform goals is to attract investments in generation capacities (at the third stage of restructuring). Thus, the following can be considered as the main officially stated goals of the electric power industry reform:

1. Ensuring energy security of the country
2. Providing sustainable functioning and development of the economy and social sphere
3. Ensuring continuous and reliable operation of power industry itself
4. Enhancing electricity production and consumption efficiency
5. Attracting investments in electricity generation

The *decrease in electricity price*, which was the main goal in the Western countries, is not mentioned (electricity cost minimization can be understood as a decrease in production cost, i.e., enhancement of production efficiency). Hence, from the very beginning it was clear that the electric power industry reform in Russia will lead to an electricity *price rise*. This means an increase in prices of all Russian goods and decrease in their competitiveness in the world markets, inflation, etc. Therefore, the reform will not facilitate the achievement of the second goal, which is related to the economy and social sphere.

The first and third goals can be mainly achieved through the *deficit-free expansion of UPS and its reliable operation*. Meanwhile, as discussed in Chaps. 4 and 6, problems and difficulties arise just here when the transition to the competitive electricity markets is made. The price barrier emerges for the entry of new producers into the market. This creates a threat of generation capacity shortage. Simultaneously, the probability of large-scale blackouts increases and the general reliability of power supply decreases, which is seen from the experience of many countries including Russia. Hence, the reform implementation will not lead to achievement of these two goals.

Stating the *attraction of investments* in generation capacities as a reform goal can be considered as a misunderstanding. On the one hand, these investments are well provided in the regulated markets through the investment components included in the consumer tariffs. As was already mentioned the Western countries even faced "overinvestment." The fact that it was not done in Russia in 2000–2006 should be considered as a serious mistake. On the other hand, attraction of private investments under the competitive wholesale market requires very high prices (4–6 ¢/kWh). With the wholesale prices of 1.5–2.0 ¢/kWh that were in the European section of UPS at the beginning of the reform there naturally could not be any private investments. However, the price rise to the "investment" level will be too costly for the economy, social sphere, and population of the country.

As to the *enhancement of electricity production efficiency* (goal 4), it can really be achieved owing to competition. However, as was shown before, this will be beneficial only for the producers whereas consumers will suffer losses due to increase in the wholesale prices to the level of marginal ones. This contradicts the second goal (ensuring reliable operation and development of the economy and social sphere). Additionally, as was indicated in Sect. 4.2, the production efficiency can also be enhanced in the regulated electricity markets with consumer tariffs established (fixed) for a long period of time (several years).

Thus, it is seen that *none* of the officially stated goals of Russia's power industry reform *will be achieved*. This means that the concept of restructuring was adopted without appropriate feasibility analysis (without comparison of the expected competition effect with the costs of organization of the competitive electricity markets and consequences of their introduction) and criticized not without reason. Or the reform initiators had the goals other than those officially declared.

In order to analyze the reform after adoption of the Law "On the electric power industry," it is appropriate on the one hand to consider the process of restructuring and on the other hand—the state of the power industry itself. As was already mentioned a new reform with a hope to attract private investments after introduction of the competitive market in fact delayed recovery of the power industry from the crisis by many years and thus exacerbated it.

The *process of reform* appeared to be difficult, costly, and dragged on. Transition period ended neither in 2005 nor in 2006. The key problem, in the author's opinion, was *inevitable increase in the wholesale electricity prices* with termination of their regulation. In the initial concept of reforms (in the Law [97]), this problem was supposed to be solved (the period of price rise to be extended) by separating and gradually expanding the unregulated trading sector in FOREM. However the prices formed in this sector were naturally lower than in the regulated sector (otherwise the buyers would not go to the unregulated trading sector). In these conditions, deregulation of prices in the wholesale market would have led to their jumping increase (by 30% or higher). *The Government did not dare to do that.* Therefore the concept of the wholesale market in the transition period had to be changed and all efforts and expenditures for the creation of the unregulated trading sector turned out to be in vain.

A new concept of the wholesale electricity and energy market (NOREM) was urgently developed by RAO EES Rossii and after several delays came into force on September 1, 2006, by the Decree of the RF Government No. 529 of August 31, 2006 [98]. Its main goal (though not declared) was also to extend the period of price increase in the wholesale market from the level of average weighted to marginal costs of generation. The new concept supposed *forced reduction* of a *share of regulated* bilateral contracts between producers and consumers from 100 to 0% during several years.

The concept of NOREM is extremely complicated. Initially *all the wholesale trade* is carried out on the basis of regulated bilateral contracts and every buyer is assigned to several producers. Producers (expensive and cheap) are distributed to every buyer so that the average wholesale price of the latter is the same as the tariff before the introduction of NOREM. Hence, every producer and buyer should conclude packages of regulated contracts in which the volumes of deliveries are calculated by the Trading System Administrator and prices (tariffs) are established by the Federal Tariff Service. Further the volumes of deliveries under the regulated contracts will decrease twice a year. By the end of 2010 these volumes will be brought to zero.

Simultaneously spot markets are organized: a day-ahead market (DAM) and a balancing market (BM). Bids in DAM should be submitted daily by all producers and buyers, and prices are formed on a marginal cost basis. Bids in BM are submitted only by those participants whose actual consumption or electricity production appears to be different from that indicated in the bids submitted in DAM on the previous day. The BM prices are also formed according to the marginal principle.

Recall that in 2001 Great Britain had already eliminated the day-ahead market, and in the balancing market electricity is traded on a pay-as-bid but not on a marginal cost basis.

The concept also suggests trading under unregulated bilateral contracts, organizing auxiliary services markets, markets for capacity (since 2008), markets for derivatives, etc. After its complete implementation, NOREM will apparently become the most complicated (and "intricate") electricity market in the world. And all that is for the sake of price rise and superprofits to be gained by the electricity producers.

The prices in the spot markets of NOREM exceed the average tariffs in the regulated contracts already now (unlike the prices that were formed in the previous unregulated trading sector in FOREM). Even larger increase will occur in the next years as the share of regulated contracts shrinks and the shortage of generation capacities appears (next section will discuss this in more detail).

It should be noted that the criticism of the reforms in Russia's electric power industry that started during the discussions of the concept of restructuring RAO EES Russii and the Law "On the electric power industry" continued during implementation of the reforms. The number of both proponents and opponents of transition to the competitive market is great. For example, the authors of [99–103] follow the officially adopted concept. At the same time the authors of [21, 58, 59, 104, 105] present the analysis of thc concept drawbacks and expected negative consequences, and point out the need for its adjustment.

As to the *state of Russia's electric power industry* itself, it grew increasingly worse. Commissioning of new capacities after 1998 on the average was below 1 GW a year. Equipment of the operating power plants was updated and modernized in the volumes that were four to five times less than necessary. The same situation was in the electric networks. As a result, after 2006 wear and tear of fixed assets reached 57.8%, that of generation equipment—62%.

Meanwhile, the managers of RAO EES Rossii did business (along with power industry restructuring): acquired electric power plants and networks in the CIS countries (Georgia, Armenia, Moldova), constructed hydropower plants in Tajikistan, paid dividends to shareholders, established high wages to managerial staff, etc. The share of the item "other expenses" in the average tariffs of the holding RAO EES of Rossii made up 47.5% in 1998 and 49.1% in 1999 [106] (author did not find more recent data). These expenses exceeded the costs of fuel, remuneration of labor, and depreciation all put together. In [104] the author made a generalized analysis of the expenses of RAO EES of Rossii that are not related to electricity production on the territory of the country.

The situation continued until the Moscow blackout in May 2005 and electricity shortage in the subsequent winter in Moscow, Saint-Petersburg, and some other regions. The administration of RAO eventually understood that energy equipment and electricity consumption decline will not last forever. The contract for purchase of thermal power plants in Bulgaria was cancelled, shareholders were asked to refuse from their dividends of the past year, etc.

The RAO administration started to urgently develop plans for the UPS expansion and investment programs for the newly created wholesale and territorial generation companies (WGCs and TGCs). The shares of WGCs and TGCs were emitted to finance their investment programs. The general scheme of electric power objects placement up to 2020 [66] was approved by the RF Government in February 2008. All this had to be done 8–10 years earlier when the volumes of worn equipment were much smaller and construction basis, machine-building plants, and design organizations were in much better conditions. Now the volumes of capital construction have increased manifold and the possibility for implementation of the developed "general scheme" is rather doubtful. The shortage of generation capacities that has already manifested itself in some regions threatens to become ubiquitous in the nearest future.

At the same time RAO EES Rossii completed its restructuring and on July 1, 2008, ceased to exist. It left the power industry split into hundreds of companies and huge plans for construction and investment programs that will have to be funded and implemented by someone else. The nearest years will show the real "merits" of top managers of RAO EES Rossii that took over the company in 1998.

Box 23 Restructuring and Status of Electric Power Industry in Russia in the Early Twenty-First Century

1. In December 2000 RAO EES Rossii submitted for approval by the Government of the RF the concept of restructuring RAO EES Rossii that would provide for transition to a *competitive* market in the industry. The Concept was thoroughly discussed and criticized. About 10 alternative conceptions were proposed. However, the Government of the RF approved "The Main Directions of Electric Power Industry Reform" by Decree No. 526 of July 11, 2001, which virtually completely coincided with the concept of restructuring RAO. This Decree initiated a new stage of reform.
2. In February 2003 after the debates that went on for more than a year the State Duma adopted the Law "On the electric power industry" that was also based on the concept of restructuring RAO EES Rossii. Some changes and supplements were aimed primarily at strengthening the role of the State and Government in the reform. In the Law there was a *transition period*, the end of which was planned on July 1, 2005, not earlier.
3. Analysis of the goals of reform that were officially included in Decree No. 526 and the Law "On the electric power industry" (as discussed in Chaps. 4–7) has shown that *not a single* stated goal will be actually achieved. This relates to ensuring energy security of the country and stable operation and development of the economy and social sphere, to attracting investments in the area of electricity generation, etc. The problem of price reduction that was the main goal in the West is not mentioned, i.e., the reform initiators understood that it would cause rise in electricity prices.
4. The process of reform proved to be difficult, expensive, and long. The transition period was not over either in 2005 or 2006. An inevitable rise of the wholesale electricity prices was the main problem caused by their deregulation. A new concept of the wholesale power market (NOREM) was urgently worked out and came into effect on September 1, 2006. It provides for conversion of all electricity trade into regulated bilateral contracts, formation of spot markets, etc. The share of regulated contracts will

be gradually *forcedly* reduced and then brought to zero by the end of 2010. Therefore the price rise will last for several years.

5. At the same time the state of the industry continued degrading. After 1998 the annual commissioning of new capacities averaged 1 GW. The energy equipment continued to wear and get out of date. This continued until Moscow blackout in May 2005 that initiated elaboration of plans for updating and construction, investment programs, etc. The time, however, was lost, the volumes of work increased manifold. Therefore, the possibility to implement these plans and programs causes doubt, in particular due to degradation of the construction complex of the industry, energy-machine building and design organizations. The electricity and capacity shortage observed in several regions threatens to become common in the nearest future.
6. On July 1, 2008, RAO EES Rossii, after completion of its restructuring, ceased to exist, leaving electric power industry unbundled in hundreds of companies, huge plans of construction and investment programs which should be financed and implemented by somebody else.

8.3 Forecast for the Years 2010–2020

The main problem in Russia's electric power industry in the forthcoming period is *shortage of generation capacities* (its prevention or overcoming). Its consequences are diverse: limited economic development of the country, electricity price rise, interruptions of electricity supply, etc. Price rise in NOREM (unjustified, in the author's opinion) occurs even now, however with the shortage to appear it will sharply increase. A high share of worn-out equipment at power plants and in electric networks makes it impossible to avoid emergencies as well.[1] Besides, capacity shortage will give rise to the "planned" limitations of consumers, in particular "rolling" ones. Shortage prevention or elimination will require corresponding measures to be taken by the Government, including electricity price regulation.

We will not endeavor to make a detailed forecast of the national power industry state and development for 2010–2020 and limit to the analysis of only two interrelated factors: possible *dynamics of prices in the wholesale* electricity market and *sources or mechanisms of financing new* generation capacities. We will try to show what the wholesale prices will be at complete transition to the *competitive* market after 2010 and what they might be under market *regulation*.

[1] The unprecedented crash of the Sayano-Shushenskaya HPP, the biggest in Russia, on August 17, 2009, is a vivid example.

Table 8.2 Indices of power industry development in Russia ([66], the base scenario with rounding)

Index	2005	2010	2011–2015	2015	2016–2020	2020
1. Electricity consumption (TWh)	940.7	1,200	–	1,400	–	1,750
2. Installed capacity (GW)	219	250	–	290	–	340
3. Its increase for 5-year period (GW)	–	–	40	–	50	–
4. Required commissioning of power plants[a] (GW)	–	–	72.4	–	67.4	–
5. Demand for capital investments[b]						
Billion rubles of 2005	–	450[d]	2,500	600	2,800	650
Billion dollars of 2005[c]	–	17.0	95	22.5	105	24.5
6. Investment component[b], ¢/kWh	–	1.4	–	1.6	–	1.4

[a] With consideration of dismantling of obsolete power plants
[b] Only for generation sphere (without electric networks)
[c] At the rate of 26.5 rubles/$
[d] Based on the expert estimation

The general scheme [66] will again be taken as the base. Although many experts (including the author) consider it overstated in the basic parameters and unrealistic, the situation will be even easier, if virtually the lower volumes of construction will be needed. The "base" (low) scenario of two that were developed in the general scheme will be considered here. The electricity consumption is assumed to grow at a rate of 4.1%. The main parameters of power industry development up to 2020 in this scenario correspond approximately to CV-3 scenario (accelerated development) discussed in [107], where the period up to 2030 is dealt with.

Table 8.2 presents indices (with rounding) of our concern for the *whole country* that are consistent with the base scenario of the general scheme. The year 2010 and the last years of two 5-year periods (2015 and 2020) are assumed to be "reference" ones for the analysis of prices. Demand for capital investments takes into consideration a generation sphere that is only responsible for wholesale prices. Investments in electric networks are supposed to increase prices for end consumers as compared to the wholesale price.

Demands for investments in the approved general scheme are indicated in rubles of the current (future) years. In the draft general scheme, however, they are given in 2005 rubles, which is more convenient. Therefore, the figures in fifth row for the 5-year periods 2011–2015 and 2016–2020 (2,500 and 2,800 billion rubles) are borrowed from the draft general scheme. For the reference years 2015 and 2020, they are assumed somewhat higher than on the average at the previous 5-year period (normally construction volumes increase by the end of the 5-year period). Demand for investments for the "reference" 2010 year is determined based on the expert estimation (it may be treated both as the end of the previous 5-year period and close to the beginning of the next 5-year period).

As for the year 2010, the situation is somewhat special. First, a small share of *regulated* bilateral contracts in NOREM will still remain. Second, all generating companies (WGCs and TGCs) have investment programs till 2012, which according to plans will be implemented basically by the means from issue of their shares, i.e., a source of financing new power plants till 2012 seems to be known. The main problems in financing in the *competitive* market will arise after 2012, when it is projected to attract *private* (some other) investments. The investment sources in the general scheme are indicated very briefly and in most general terms—"own funds of generating companies (depreciation, profit on capital investments, accumulated profit of past years, funds from VAT reimbursement) and attracted funds (credits and issue of shares and bonds)." It may happen that the funds from issue of shares will be insufficient to implement investment programs of WGCs and TGCs till 2012 and there will be no funds to finance further construction. Therefore, the Government would probably have to take regulating measures even in 2010–2012.[2]

For the performed calculations to be comparable with the results obtained in Chaps. 5 and 6, capital investments are converted into US $ of the year 2005 (at the rate 26.5 rubles/$). Thus, all calculations were made in fixed 2005 prices expressed in dollars (or cents).

The last row (for the "reference" years) of Table 8.2 presents an investment component that is necessary for updating and expansion of generation capacities, if the latter is performed by "self-financing" (mechanism 1 addressed in Chap. 6). This corresponds to the variant of our forecast, when electricity price *regulation* is introduced and the State takes measures to provide required commissioning of power plants. This investment component is supposed to be included in *all the electricity* consumed in the country. It is determined, therefore, by division of the required investments (fifth row, in dollars) by electricity consumption (first row).

Note that roughly the same values of the investment component will be needed, provided construction is financed owing to bank credits or private investments in the single-buyer market (mechanism 2). In Chap. 6 it was shown that mechanisms 1 and 2 of financing are equivalent, if the development pace λ equals the interest on capital σ. Although in the base scenario of the general scheme the rate of electricity consumption growth was assumed to be 4.1%, generation capacities should be commissioned (constructed) at a higher pace ($\lambda = 7$–10%), if account is taken of dismantling the outdated power plants. The interest on capital σ will be almost the same at the *guaranteed* repayment of credits or recovery of investments. Calculations of the investment component under mechanism 2, if applied to the general scheme, would be highly complicated and they were not made. Because of the mentioned reason, however, the values of the investment component (sixth row in Table 8.2) will be assumed to correspond to the *variant of introduction of the state regulation regardless* of the fact whether investments in generation capacities are included directly in consumer tariffs or bank credits and private investments are used.

[2] The 2008 world financial crisis (turning into the economic one) had a major impact on power industry development in Russia. It complicated implementation of the investment programs of WGCs and TGCs, in particular receipt of credits. Simultaneously a new decline in electricity consumption will occur and it will again facilitate a current situation, but aggravate it for the subsequent period. Now it is difficult to assess all consequences of the world crisis and we will not try to do this in the book. It will obviously delay the overcoming of the own crisis in power industry of the country. The general picture of wholesale price formation under regulation and deregulation, however, will correspond to that described in this section.

Comparison of the values of the investment component for 2010 and 2015 in Table 8.2 with their values in Table 8.1 shows that they increased almost twofold. Actually the increase is more sizable since in Table 8.1 the investment component was determined including capital investments in electric networks but in Table 8.2 only in generation capacities. This is explained by the fact—delay ("shift") in launching a *large-scale* process of power industry re-equipment and development in Russia almost for 10 years.

The value of the investment component (1.4–1.6 ¢/kWh) is very high. The tariff rise for end consumers will be still higher due to investments in electric networks. At the same time in the variant of *competitive market conservation,* rise of the wholesale prices for the private investments to be recovered will be still more sizable.

Figure 8.2 presents costs (taken from Fig. 6.2 in Chap. 6) of new power plants. These costs correspond to prices that should be in the wholesale market of the European section of Russia's UPS for construction of new power plants in the variant of competitive market conservation. The dashed lines in the figure indicate

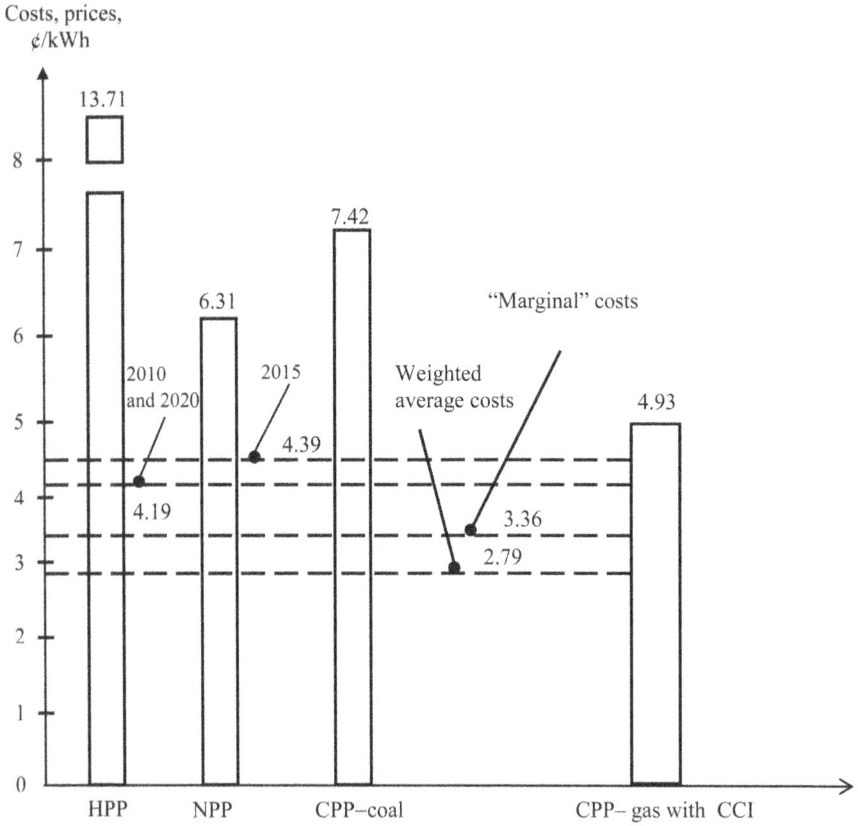

Fig. 8.2 Prices of the wholesale market of Russia's EUPS in 2010–2020 under regulation and competitive market

weighted average costs in the generation sphere of EUPS of Russia at 2010 level (2.79 ¢/kWh) and the "marginal" costs (CPPs on gas with steam turbine installations), to which the wholesale prices will rise after they become deregulated (3.36 ¢/kWh).

Further it was assumed that the weighted average costs of operating power plants (at fixed 2005 prices) will remain the same (2.79 ¢/kWh) in 2015–2020, though actually they can vary due to general increase in electricity consumption and changes in the structure of generation capacities in EUPS. Hence, the investment component from Table 8.2 is added to these costs and in Fig. 8.2 the dashed lines show tariffs of the wholesale market of EUPS that will be in the reference years in the *price regulation variant* (1.19 and 1.39 ¢/kWh)

It can be seen that in the competitive market even construction of gas-fired CPPs with combined-cycle installations requires still higher prices. In order to attract private investments in NPPs and coal-fired CPPs, the prices should be 2–3 ¢/kWh higher than the regulated ones. As for HPPs, their construction under the competitive (completely unregulated) market should be admitted impossible in general. Surely, for concrete power plants and conditions of financing their construction the figures may differ from those given in Fig. 8.2; however, the basic picture will be such as is shown there.

Note that high prices to attract investments in new power plants in the competitive market are *objectively necessary*. The situation cannot be "saved" by either the portfolio investments at the issue of shares by the generating companies or newly organized capacity market. The portfolio investments (if they are not spent for dividend payments to former shareholders) at insufficiently high prices of the wholesale market can result in their lack for construction of new power plants in the planned volume or impossibility of their recovery by output of these power plants and the company will find itself in a difficult financial situation or even become bankrupt. As to the capacity market, the prices there will be very high, which will lead to price soar in the wholesale electricity market as well.

Rise in the wholesale prices to the "investment" level will lead, as mentioned in Chap. 6, to getting the *monopoly profit* by *operating* producers and naturally to unjustified consumer expenditures (and also to inflation, etc.). *Only new* producers are allowed to have high prices and it can be achieved only in the *regulated* markets organized according to Model 1 or 2.

A general level of wholesale prices even at their regulation (4.0–4.5 ¢/kWh) is very high for Russia. They can be decreased by *using* the *Stabilization Fund* to finance construction of power plants and networks planned in the general scheme [66]. Shortage of generation capacities and price rise destabilize the national economy and will delay its development. To prevent this situation requires investments of hundreds of billion dollars and utilization of the Stabilization Fund for this purpose seems to be quite reasonable and wise. The Fund has been accumulated by the whole country and the benefit from uninterruptible and cheaper power supply will also be gained by the whole country.

In conclusion of this section, tentatively the following generalized *forecast for Russia's power industry* can be made for the coming decade:

- Constant and very sizable price rise in the wholesale electricity market (NOREM) till 2010 may be expected because of two reasons: (1) decrease in the share of regulated bilateral contracts and (2) capacity and electricity shortage available in the country even now. This price soar cannot be treated as justified (appropriate). It will result in extra profits (producers' surplus) to power generating companies and cause damage to all electricity consumers.
- New NPPs, CPPs on coal, and particularly HPPs will hardly be constructed by private investors. To do this, the prices should additionally be raised to the level completely unacceptable for the economy and population of the country and besides, the operating (already existing) producers would receive huge monopoly profits. This fact will aggravate capacity shortage and will demand taking "non-market" measures by the State to finance generation capacity expansion.
- Uncontrolled price rise in NOREM (particularly, at shortage toughening) that is accompanied by extra profits of generating companies will prove to be inadmissible for the country even in the coming years and the State will have to restore electricity price regulation.
- Return (with corresponding adjustments) to the two-level structure of regulated markets that was created in the 1990s with improvement of regulation mechanisms (methods, procedures) is most expedient. Such a two-level structure seems to be the best for Russia. When perfecting the power industry structure and management, it is advisable to use the experience of China, USA states with regulation, France, Japan, and Brazil.

Return to the two-level structure of regulated markets will not be very complex, as may seem at a glance. It will not require change of WGC property, but only the introduction of regulation in their tariffs. The federal network company may be combined with the System Operator. In this case it will perform functions of the Purchasing Agency and include the Trading System Administrator that was converted into department to carry out mutual financial settlements.

Regional power systems, in particular with territorial generating companies, will require more complicated transformations. These TGCs consisting of CGPPs are a "creation" of Russian reformers and have no analogs in other countries. It is quite probable that their organization can prove to be erroneous because of several reasons. First, many TGCs will be monopolists in their territories in view of their high share in generation capacities of the corresponding zones (nodes) of the wholesale market. Second, their interaction with regulated heat supply systems in cities including individual CGPPs of each TGC is not clear. The regional or municipal bodies that regulate tariffs for heat energy will tend to fix them at a minimum possible level (individual for each CGPP). The conflicts between managers of TGCs and regulatory bodies are inevitable. Whether CGPPs will be able to be competitive in the *electricity* market and cover their *full* costs at the regulated tariffs for *heat energy*, depends on the results of settling these conflicts. Third, it is not clear what funds will be used for TGC expansion (construction of new CGPPs).

Because of the indicated reasons, the question about disintegration of TGCs can arise during continuing transition to the competitive market. This will be an additional argument to restore regulated vertically integrated companies at a regional level and simplify return to them.

In principle, the issues of restoring the state electricity price regulation, determination of the proper market structure, and provision of deficit-free and optimal industry development in Russia obviously call for special and besides urgent studies. One of the author's goals of the book is encouragement of such studies.

Box 24 Analysis of Conditions and Problems of Russian Electric Power Industry Development up to 2020

1. The key problem in the future period is *prevention of generation capacity shortage*. It requires investments and capacity commissioning that should be higher by ten times and more than those in the recent years. It seems practically unreal and the fact of shortage has to be accepted as inevitable.
2. In the case of competitive market, the shortage will lead to multiple rise of prices in the wholesale electricity market, which will be inadmissible for the economy and social sphere of the country. Therefore, the Government of the RF is likely to *introduce price regulation*.
3. Price regulation will cause the problem of attracting *private* investments in new power plants that are possible only at very high prices. Hence, along with the price regulation the other sources of financing UPS expansion will be needed.
4. An *investment component* of consumer tariffs can be one of such sources. The calculations performed have shown that in this case the wholesale prices will be lower than in the competitive market, but still very high because of delay of the *large-scale process* of equipment updating and further development of the industry almost for 10 years. Under these conditions it is advisable to *take advantage of the Stabilization Fund* to finance construction of power-producing facilities.
5. On the whole, in the coming 5–8 years the competitive market in Russia might be expected to suffer failure and the state regulation in electric power industry is likely to be restored. Then it is reasonable *to return* (with proper adjustments) to the *two-level structure* of regulated markets of the 1990s with *improvement* of the regulation methodology.
6. The problems and concrete ways to ensure deficit-free and optimal expansion of UPS of Russia, state regulation of electricity prices, and expedient reform of the electricity market call for further and urgent studies.

Chapter 9
Conclusion: Main Results and Directions for Further Research

9.1 Relatively New Results Obtained in the Book

Without pretence to the total novelty, the author finds it necessary to make an emphasis on the following results presented in the book:

1. *Studies on the EPS properties and their influence on the electricity market.* The fact that the market is created in a very complicated and capital-intensive electric power system characterized by particular properties has conditioned an *extreme imperfection* of this market and its *principal differences* from the markets in other industries. Analysis of these properties revealed the following distinctions of the industry and its markets:

- *Economies of scale* are characteristic of the EPS as a system. The *EPS integrates the effects* obtained owing to technological progress and other measures in electricity generation, transportation, and distribution. The "capacity" effect of EPS interconnection is very important. It implies a decrease in the required capacity of power plants when the intersystem transmission is constructed. The experience of EPS expansion in all the countries and the formation of National Power Systems and interstate power pools in the second half of the twentieth century proves the *existence and permanent manifestation* of economies of scale. In Sect. 2.4 the quantitative estimates of the effect of creating the UPS of the USSR are presented.
  The statements that power systems have lost this effect and power industry has ceased to be a natural monopoly with the advent of highly efficient combined-cycle plants can only be explained by insufficient knowledge of the power system design methodology and the methods for efficiency estimation of intersystem tie lines.
- *The existence of a physical barrier* for new power producers to enter the market *in the short run.* Thus, one of the basic conditions for *perfect* competition is not met. The physical barrier that fences electricity market cannot be overcome by any methods and means. Therefore, the attempts to organize a *competitive* electricity market (that supposes perfect competition) should be considered as *contradicting* the theory of microeconomics.

L. S. Belyaev, *Electricity Market Reforms*,
DOI 10.1007/978-1-4419-5612-5_9, © Springer Science+Business Media, LLC 2011

- *The emergence of a price barrier* under the competitive market for new producers *in the long run*. Here the dilemma (contradiction) occurs:
  - Either with the wholesale market prices corresponding to the costs of *operating* power plants the *new* power plants will not be constructed and this will cause capacity and electricity shortage.
  - Or the prices have to be increased to the level at which the investments into *new* power plants will be paid back and *operating* producers will get *monopoly* profits paid by consumers. This level is relatively low in the countries capable of constructing new power plants with gas turbine and combined-cycle installations on cheap natural gas.

    For cases that require the construction of capital intensive HPPs, NPPs, or coal-fired CPPs, the contradiction can be resolved only with the *state regulation* of electricity prices and EPS expansion. High prices that are required to pay back the investments should be obtained by *new* producers *only*.

- *Principal difference between instantaneous (hourly) costs* of the electric power plants that are used to optimize EPS operation and short-*run (yearly) costs* that determine the real value and price of electricity. There is no such difference in other industries and the neglect of this feature of EPS has led to the attempts to organize electricity *spot* markets with real-time trade (with hourly or half-hourly intervals). Meanwhile hourly cost characteristics of the power plants reflect only *variable* costs (do not include fixed costs) and may not be used to determine electricity prices. The latter are formed on the basis of *total* costs over the *short-term* period (a year) as a whole. Therefore the spot market organization as is shown in Sects. 5.1 and 5.2 *contradicts the theory of microeconomics*. Practical experience of spot market operation confirms the erroneousness of their creation. The day-ahead spot market has been abolished in Great Britain and Brazil.

  Electricity can only be traded through *long-term* contracts (1–3 years) that stipulate the prices reflecting the *total* short-run production costs (including fixed costs).
- *The need to design and plan the EPS expansion as a unity on a centralized basis.* As any technical system EPS should be optimally designed and expand. This, seemingly, quite an obvious circumstance is neglected by the developers of the *competitive* electricity market conceptions, though the need for *centralized control* of EPS *operation* is recognized by everybody. The nonmarket measures that have been taken lately to ensure expansion of generation capacities in EPS, for example, capacity markets, at best can prevent electricity shortage but cannot guarantee an *optimal* mix of power plants in EPS. The latter is possible only under the centralized planning of EPS expansion.

2. *Studies on electricity generation costs in the short and long run.* Essential differences were revealed between the curves of average costs of power plants, the generation sphere of VICs and PGCs, and the costs of "typical" firms considered in microeconomics:

- *In the short run,* the relationships between *average variable and the total* costs of *power plants* and the annual electricity *output* do not have an U-shaped form (with minimum). Average *variable* costs (and, hence, *marginal* costs) virtually for all kinds of power plants remain constant and average *total* costs reach the minimum at the maximum annual output always *exceeding* the marginal costs. Thus, power plants should participate in the *competitive* wholesale market with their *total* (but not marginal, as it is accepted in the theory of microeconomics) costs not to be a bankrupt.

  Besides, the *short-run* (annual) *variable* costs of power plants are to a great extent *uncertain* as they depend on EPS operation throughout a year, which is optimized centrally in accordance with daily and seasonal load changes and mix of operating equipment at all power plants. The *total annual* costs of each power plant will also be uncertain.

  Taking into account the features presented here and above,the *short-run* electricity market should be organized on the basis of the following conditions:

  - *Long-term* contracts should be used for trade; prices have to correspond to the *average total* annual costs (and not variable hourly costs as it happens in the real-time spot markets).
  - The price offers of power plants in the *competitive* wholesale market should include their *total* costs (and not marginal as it is assumed for "typical" firms).
  - The *uncertainty of total annual* costs of power plants should be taken into account one way or another.

  In the regulated markets (Models 1 and 2) these conditions are naturally met, whereas in the competitive markets special measures have to be taken.
- In the generation sphere of vertically integrated company, the costs of its power plants are *averaged.* Under state regulation, these average weighted total costs are included in the tariffs for consumers (along with the costs of electricity transport, distribution, and sales). Similar averaging occurs under the regulated single-buyer market. Therefore with electricity market organized according to Models 1 and 2, the regulated prices (tariffs) for consumers include *average weighted* costs of VIC (or EPS) generation.

  In the *competitive wholesale market* (Models 3 and 4) where electricity is supplied by several independent (nonregulated) PGCs, the prices will be formed at the level of total costs of the *most expensive* (the least efficient) marginal power plant in the EPS. These prices will be considerably higher (for European section of the UPS of Russia by almost 30%) than the average weighted costs of VIC (or EPS) generation. More efficient power plants in all PGCs will receive superprofit (producer's surplus) paid by consumers. This superprofit is in no way connected with the enhancement of production efficiency (and is not a desert of producers) and is obtained exclusively owing to the property (in this case a negative one) of the market with free prices.
- *In the long run,* the investment component necessary for the expansion of generation capacities is added to the direct (operation and maintenance) electricity

generation costs. An individual *operating* power plant may have no investment component if the investments have already been recovered. Therefore in the long run, the costs of *new* power plants being designed or planned for construction or those already constructed but continue to pay back the invested capital are important.

In the regulated vertically integrated companies (Model 1), investments into new power plants are included in the investment component of tariffs for consumers and paid back at the expense of *total electricity output* of the company. As is shown in Sects. 6.1 and 6.3, similar situation (though a more complicated one) will be observed in the regulated single-buyer market (Model 2). Long-run costs of VICs and individual PGCs in the market organized according to Model 2 are in harmony with their understanding in the theory of microeconomics.

The long-run costs of generation acquire a principally new sense in the *competitive* wholesale electricity market (Models 3 and 4). Both for a new producer that constructs one (its first) power plant and for operating PGCs (see Sects. 6.1 and 6.3) the *costs of new* power plants (including an investment component) will be their long-run costs. These costs are naturally essentially higher than the costs of operating power plants. This creates the above mentioned price barrier in the long run.

3. *Studies on financing mechanisms for construction of power plants.* Three main financing mechanisms were determined (Sect. 6.1).

- *Mechanism 1*—self-financing in the regulated monopoly company (Model 1). Here the investments in the new power plants are included in the investment components of tariffs for consumers *during the period of their construction*. Further the consumers pay only operating costs of electricity generation. In this case the investment component of tariff will depend on the pace of generation capacity expansion λ (Sect. 6.2).
- *Mechanism 2*—construction of power plants at the expense of *credits* in the regulated monopoly (Model 1) or by *private investors* under the single-buyer market (Model 2). Here a power plant is constructed "free of charge" for consumers but then they pay back credits or private investments during some period T_R at an interest rate σ. This payback is included into the investment component of tariff and is paid at the expense of the *total* electricity supplied to consumers. The investment component of tariff, along with T_R and σ, will also depend on the pace of expansion λ.
- *Mechanism 3*—construction of power plants by *private investors* under the *competitive* wholesale market (Model 3 and 4). In this case the investments into each power plant will have to be paid back at the expense of electric power produced by *this one* power plant. The investment component of the price required to payback the investments will depend on the payback period T_R and the interest rate σ. The latter, due to risk for investor, will be higher than under mechanism 2 where the investment payback is guaranteed.

The analysis of mathematical expressions for investment components of tariff or price under the three financing mechanisms (Sect. 6.2) shows that:

- Mechanism 1 (self-financing) is profitable for regulated monopoly if the expansion pace λ is lower than the interest rate σ ($\lambda < \sigma$) and vice versa, and mechanism 2 (crediting) is preferable if $\lambda > \sigma$ (profitable in the sense that the investment component of tariff is lower). This means that at a fast EPS expansion pace crediting or the single-buyer model is more preferable.
- With equal payback period T_R and interest rate σ the investment component under mechanism 2 is *always lower* than under mechanism 3. Hence, in the markets with regulated prices (Models 1 and 2) the required expansion of generation capacities can always *be ensured* by lower tariff or price increase than under competitive markets (Models 3 and 4).

These trends were revealed in organization of electricity markets in different countries.

4. *Studies on efficiency of intersystem and interstate electric ties.* Transition to a competitive market causes difficulties not only in expanding generation capacities (price barrier) but in constructing transmissions as well. Two such problems are considered in Sect. 6.5.

- In the competitive market electricity *export ceases to be mutually profitable.* Export raises the demand for electricity and its prices in the exporting country, i.e., gets *unprofitable for the consumers* of this country. At the same time export increases supply and decreases prices in the importing country, which is *unprofitable for the producers* of the importing country. This will undoubtedly cause opposition to the construction of export transmission lines, complicate, or even prevent the construction. Meanwhile in the markets with regulated electricity prices it is possible to ensure export profitability for consumers of both countries and preserve profitability for their producers.
- Practical *impossibility to substantiate* financial (commercial) efficiency of *reverse* intersystem and interstate transmissions intended to implement the *capacity* effect of interconnecting power systems. This is explained by separation of electricity generation and transportation spheres under the competitive market and by change in the financing mechanisms for transmissions (sources of investments and payback terms). The situation becomes very similar to the construction of new power plants.

The described difficulties have led to a sharp decrease or even termination of the construction of intersystem and interstate electric ties in the countries with deregulated power industry.

9.2 Practical Experience of Power Industry Restructuring

The analysis of power restructuring experience in Russia and other countries that was carried out in Chaps. 7 and 8 confirms to a large extent the results of theoretical and qualitative researches described in Chaps. 2–6. The following points can be underlined.

1. *Power industry deregulation is not a global tendency.* The extent to which power industry was deregulated (if at all) and its results are country specific.

- Most countries have *preserved the regulated monopoly companies* (Model 1). In doing so they have allowed connection of independent power producers to their networks and some countries have introduced separate accounting for expenses (costs) for the spheres of electricity generation, transportation, distribution and sales. The most vivid example is the USA and Canada where most of the states or provinces did not carry out deregulation. Regulated VICs have also been preserved in France, Japan and in most of developing countries in Asia and Africa.
- Many countries *stopped* the reforms with the *Single-buyer market* (Model 2) and continued to regulate electricity prices and expansion of their EPSs. Such markets were organized in China, India, South Korea, Brazil and some other countries. They make it possible to maintain moderate prices despite high rates of electricity consumption growth, power shortage and the need to construct capital intensive power plants (HPPs, NPPs, etc.).
- In the countries *deregulating* power industry (being in transition to competitive markets according to Models 3 and 4) the process of restructuring has turned out to be hard and long. The conceptions of restructuring are revised (the reform of reforms) and in no country the process can be considered completed. Electricity markets and trade grow increasingly complicated. This is explained by many drawbacks that were revealed with termination of electricity wholesale and retail price regulation.

2. *Problems of investing into EPS expansion* are, as a rule, crucial for organization of electricity market.

- *Developing countries* virtually always suffer from the lack of investments due to high rates of electricity consumption growth and the need to attract private investors or use bank credits. These problems are most rationally tackled under *regulated* markets (Models 1 and 2):

  - Receiving credits by regulated monopolies or concluding long-term contracts with independent power producers with the prices providing investment payback at a rather high interest rate (as was done in China during 1980s–1990s).
  - Holding tenders or auctions for construction of new power plants under the single-buyer market, which is currently practiced in China and Brazil.

  In these cases high electricity prices are paid only to new power producers (financing mechanism 2 in Sects. 6.1 and 6.2). Besides, the construction of capital intensive power plants (HPPs, NPPs) is also possible. The wholesale prices can be maintained at a rather moderate level.

  Under competitive markets (Models 3 and 4) investing into generation capacities is possible only if there is cheap natural gas for gas turbine and combined-cycle plants. This was observed in Argentina and Chile. In Brazil where the share of hydropower plants is rather large construction of these plants turned out to be too risky for private investors (and they were not constructed). As a result all the three countries in some years after the introduction of competitive market faced crises caused by shortage of generation capacities:

- Brazil—in 2001 due to complete termination of new power plants construction.
- Argentina—after 2001 due to a general economic crisis, devaluation of the national currency and cessation of private investments.
- Chile—after 2004 when gas export from Argentina was limited and then stopped.

After these crises the state regulation of wholesale prices and expansion of EPS was revived and strengthened. For example Brazil, owing to efficient regulation, increased network construction which was very important for this large country.

- *In developed countries* the investment problems are different and they are solved differently depending on three main interrelated factors:

  - The model of electricity market organization.
  - Available resources and price of natural gas.
  - Relationships between the available generation capacity reserves and the rates of electricity consumption growth (at large reserves and low rates the construction of new power plants will not be topical during several years).

  In the countries (states and provinces) where regulated monopolies are preserved (Model 1), the investment problems are solved as before. There are no particular difficulties. The problems arise with deregulation.

  Transition to a competitive market leads (as is shown in Sect. 6.4) to an essential increase in the wholesale prices if there is no cheap natural gas and it is necessary to construct capital intensive power plants, particularly NPPs. Therefore, France and Japan maintain regulated monopolies and South Korea, which planned the transition, settled on the Single-buyer market.

  Countries that implemented deregulation had as a rule favorable starting conditions: large capacity reserves, well-developed electric networks, etc. Some of them had the possibilities to use natural gas.

  However, after deregulation:

  - the construction of HPPs and NPPs stopped everywhere and in some countries the construction of coal-fired CPPs ceased as well;
  - in the 1990s England and in the early twenty-first century the USA saw a boom in construction of power plants with gas-fired combined-cycle installations, and as a result Overinvesting occurred, which was previously considered a drawback of regulated monopolies only;
  - network construction declined sharply;
  - in many countries (for example, in Scandinavia) capacity reserves reached the critical level.

  The state of California and the province of Ontario faced the crises that forced them to return to regulation. These crises could partly be explained by the absence or insufficient amount of new capacities put into operation.

On the whole, the investment problems in EPS expansion should be considered as "devastating" for the competitive electricity market. With time they will lead to the crises (similar to those in Brazil and Chile) after the reserves decline to an inadmissible level, the possibilities to use natural gas exhaust or its price rises, and the need emerges to revive the construction of "traditional" capital-intensive power plants.

- In *Russia,* where the deregulation process is still in "full swing," its drawbacks in terms of investing into UPS expansion have already manifested themselves. There were no real ("project") private investments into generation capacities and there are no grounds to expect them in future. The "portfolio" investments that were obtained owing to the issue of shares of generation companies make up no more than 15% of the total demand for investments (into the generation sector of Russia's UPS) till 2020. They will apparently be insufficient even if the WGC and TGC investment programs are implemented by the year 2012. Shortage of capacities and electricity in the country is inevitable in the years to come. The 2008 world financial and economic crisis will decrease the electricity consumption and mitigate the shortage. At the same time it will complicate financing the investment programs and worsen the situation in Russia's power industry by the time it ends.

3. *Electricity prices rose under market deregulation* in many countries that have not faced electricity shortage yet for several reasons. One of them is the *loss of economies of scale*. It is very difficult to quantitatively estimate the influence of this factor, if possible at all. However, it undoubtedly reveals itself everywhere.

Two more obvious reasons—additional expenses for *organization* (*creation*) *and operation of competitive markets* and increase in *administrative and overhead costs* in many newly established companies. In principle these additional costs and related price rise can be calculated for each specific market. Their value will certainly be different but rather noticeable.

For some countries, the following reasons for electricity price rise can be indicated in addition to the three general factors:

- In those states of the USA that introduced competitive market the prices undoubtedly soared to the level of *marginal* ones corresponding to the costs of the most expensive (marginal) producer participating in the market. This also implies that consumers started to pay the producer's surplus.

  Besides, in some states the prices could increase to the level of costs of *new power plants* with combined-cycle installations on natural gas. Judging by the CCI construction boom that was observed there in the early twenty-first century this price level was lower than or close to the marginal prices based on the costs of other types of operating power plants. After a two- or threefold natural gas price increase in 2002–2004, this level has naturally risen and can become determinant. However, after a sharp drop of oil (gas) prices in 2008 the wholesale electricity prices in these states will apparently return to the level of marginal ones.

- In Norway and Sweden the price increase was considerably affected by electricity *export* (Sect. 6.5) though marginal price setting was also observed. In these two countries consumers probably bore the largest losses due to deregulation (in Western Europe).
- In Germany and other countries along with the price increase to the level of marginal costs the producer's abuse of market power, i.e., various price manipulations was observed. These manipulations occurred to a greater extent in California during the crisis of 2000–2001.
  On the whole the electricity price increase in the competitive markets as compared to its possible level under market regulation should be considered quite natural. It can be expected in all the markets where it has not occurred yet.

4. *Other consequences of electricity market deregulation* include the following:

- Decrease in power supply reliability. This, along with massive disconnection and limitations of consumers during the crises in California and Brazil in 2000–2001, is confirmed by large blackouts in the northeast part of the USA and in Western Europe in 2003, Moscow blackout in 2005, rolling blackout in Texas in 2006, etc.
- An extraordinary volatility and unpredictability of prices in the electricity spot markets, which have been observed everywhere. This has led, as was already mentioned, to rejection of the day-ahead markets in England and Brazil.
- Superprofits of power generation companies due to the wholesale electricity prices exceeding the costs of companies. This was observed in Scandinavia, England, and some other countries and resulted in particular in the enrichment of top-managers of energy companies including those in Russia.

5. *An attempt to organize a competitive market of long-term contracts* in the UK with introduction of NETA in 2001. The conception of NETA (and then BETTA) suggests creation of a *forward market* (*exchange*) for *standardized long-term contracts* for the period up to several years. We can suppose that this should be a real competitive market of long-term contracts, which is only possible in the electric power industry (described above). This market can give "*price signals*" on electricity production in the short run and on market expansion (or shrinkage) in the long run. However, as far as the author knows, this segment of BETTA market has not been organized yet. Long-term bilateral contracts are concluded outside the exchange, i.e., as individual transactions. Prices in these contracts are confidential and no price signals are generated. Hence there is no world experience in creating a *competitive market for long-term contracts,* which is required in electric power industry.

6. *Power industry reforms and their consequences in Russia* are mainly similar to those described above for other countries. The reform turned out to be hard and lengthy. The transition period end date has been shifted many times and now it is planned for the end of 2010. The conception of reform for the transition period has been radically changed. A new conception (NOREM) is extremely complicated and suggests a gradual (during several years) increase in the wholesale prices to marginal ones.

Unlike reforms in western countries Russia's reform of power industry started with low electricity prices and their decrease was not a goal of the reform. To the contrary the price increase was covertly implied. Meanwhile low prices of energy carriers (including electricity) are a benefit for Russia with its severe climate and vast territory (high transportation costs). Unjustified electricity price rise will result in the fall of Russia's economic competitiveness, deterioration in living standards, inflation, etc.

The analysis of officially declared goals of the power industry reforms, presented in Sect. 8.2, shows that in fact none of them will (can) be achieved. This concerns in particular the attraction of private investments for upgrading and expansion of generation capacities (discussed above). Even now the prices in the spot markets of NOREM are increasing rather fast. This trend will persist due to decrease in the share of regulated bilateral contracts and shortage of capacity and electricity. The electricity producers will receive higher and higher superprofits.

We can expect the RF Government to introduce regulation of electricity prices, undertake special measures to finance UPS expansion, and change the conception of power industry restructuring. In doing so it is certainly necessary to take into account the current financial and economic crisis. The author believes it will be most expedient to restore the two-level structure of regulated markets that was created in the 1990s, to make the appropriate corrections, and to improve the system of state regulation of tariffs and UPS expansion.

9.3 Analysis of Initial Principles (Arguments, Postulates) of the Competitive Electricity Market Conceptions

The basic arguments and postulates that represent the initial principles underlying the conceptions of competitive electricity markets are listed in the introduction. In this book the author has tried to show their *theoretical* groundlessness or erroneousness, which led to numerous flaws and somewhere even to a total failure of the competitive electricity market. Let us consider these principles in the order they are given in the introduction.

1. *The possibility of creating conditions for perfect competition in the wholesale and retail electricity markets*. This point can be considered basic for substantiating the termination of electricity price regulation. Speaking of the "competition efficiency" the speakers always assume perfect competition.
   However, as is shown in Sect. 3.2, almost none of the conditions for perfect competition are satisfied in electric power industry. This, incidentally, is recognized by everybody—the author has not met statements that electricity market is perfect. However, the supporters of competitive market think that it *can be made* perfect.

Without detailed consideration of possibilities for creation of all the required conditions for perfect competition in electricity market, we will only note the practical *impossibility of ensuring* one of the main conditions—*free* entry of *new firms into the industry*. As was already mentioned, *in the short run* there is the *physical barrier* for new power producers, i.e., a power plant should be constructed and connected to power system and this requires several years. No measures on "appropriate designing" of the market can help overcome this physical barrier and make electricity market perfect. In *the long run*, as is shown in Chap. 6, under the competitive market new power producers face a *price barrier*, i.e., their free entry is also impossible. Operating producers possess market power both in the short and long run.

Hence, the presence of physical and price barriers for NPPs makes it impossible to create conditions for perfect competition in power industry. Naturally, termination of price regulation leads to troubles.

2. *Modern power systems have lost economies of scale and vertically integrated energy companies have ceased to be natural monopolies.* The erroneousness of this argument has been discussed above.
3. *State regulation cannot be made effective.* This point was not considered in detail in the book. In Sect. 4.2 the author noted only the possibility to create incentives for power producers to enhance the production efficiency through establishing tariffs for a long period of time (several years). However, there are theories concerning regulation of natural monopolies (mentioned in [58]). Electricity markets are regulated in many countries and their system (methodology) of regulation is undoubtedly improved. Therefore this statement can by no means be recognized as decisive for transition to nonregulated electricity markets. If regulation is necessary it should be improved but not rejected because of difficulties it causes.
4. *Competition in the wholesale market will decrease wholesale electricity prices.* As is shown in Chap. 5 and is generally well-known in the theory of microeconomics liberation of wholesale prices, vice versa, causes their increase to the level of marginal (most expensive) producer. Thus more efficient producers get superprofit (producer's surplus), which is paid by electricity consumers.

   The hopes for price decline "in the future" are as a rule not supported by quantitative calculations (the author has not found such calculations). The available facts of wholesale price decrease after introduction of competitive markets (for example, in Chile and England) were related to many circumstances. Along with competition there were other reasons (a general increase in the EPS operation level, conversion to natural gas, a decrease in fuel prices, etc.) which could also lead to decline in the wholesale prices under regulation. This is proved, for example, by a jump in the profits of generation companies after deregulation in England.
5. *It is possible to organize an electricity spot market with real time trade.* In Sects. 5.1 and 5.2 the author shows the incompatibility of spot markets (a day-ahead market) in the context of the theory of microeconomics (this was already spoken of above). Their flaws manifested themselves in practice as well.

Therefore electricity spot markets no longer exist in some countries (England, Brazil).

6. *The market itself (without regulation) can provide the required expansion of generation capacities in EPS.* The erroneousness of this point is shown in Chap. 6. Practically it is confirmed, on the one hand, by numerous cases of power shortage (crises in California, Brazil, Ontario, Chile, etc.) and, on the other hand, by overinvesting that was observed in England and the USA. This can be avoided only with centralized designing and planning of EPS expansion as a single system.
7. *Creation of retail electricity markets is important to ensure "the right of consumers to choose a supplier".* It seems that this "right" is artificially invented to justify deregulation of prices in the retail markets since the real effect of competition in these markets is obviously lower than the expenditures for their organization and operation. In fact the situation is that the consumers will get the right of choice but the prices of each supplier will be higher than the previous prices of one regulated monopoly supplier. The retail electricity markets were not organized in Chile and Brazil.

   Generally most of the considered arguments for organization of the competitive electricity market are theoretically groundless while the rest of them cannot be sufficient causes for termination of price regulation.

9.4 General Conclusions

1. *Deregulation of electric power industry* (transition to a competitive market) *should be recognized as a mistake.* The conceptions of competitive electricity market are insufficiently thought through and substantiated. They do not take into account the important properties and features of electric power systems. Many of their points contradict the theory of microeconomics and some of them are simply declarative. It is quite logical and natural that practical implementation of competitive markets revealed many flaws. Moreover in some economies they resulted in severe consequences (crises) that generated the need to restore state regulation of electricity prices.

2. *There are many causes of deregulation* (this mistake) and their combinations are different in different countries:

- *A wave of passion* for market, competition, liberalization, privatization, etc. in the late twentieth century. This affected the frame of mind of the public including scientific, industrial and governmental circles. In the last decade this wave has diminished which is shown, in particular, in the works by the Nobel prize winner Professor Joseph Stiglitz [51, etc.]. The financial (and economic) crisis that started in 2008 will "cool" this passion even more.
- *Delusions and insufficient professionalism* of developers of the original competitive market conceptions. This manifested itself partly in the insufficient knowledge and consideration of the features of electric power industry and power

systems, and partly in the knowledge of the theory of microeconomics. At the same time they can hardly be reproached for that since the electricity market problems are very complicated and peculiar and were very little studied at that time and moreover there was no practical experience.

- *Interest of power producers in deregulation.* Though there were cases where monopoly power companies (especially private ones) objected to their splitting, still most of producers understood the benefit of deregulation and by all means facilitated it. One of the vivid examples is Russia, where RAO "EES Rossii" was an initiator and most active implementer of the reform.
- *Interests of governments* in some countries (England, Brazil, Norway, etc.) which implied either budget increase through privatization of power plants or increase in tax receipts owing to electricity price rise. Certainly these "interests" do not correspond to the true interests of the state, economy, and population of the country; however, they did occur. The governing bodies of the European Union are showing extraordinary persistence in deregulation of power industries of member states and creation of a single European electricity market and hardly understand the erroneousness of their actions.
- *Following the example of other countries* not making proper account of specific features of their own country. First of all, this concerns the countries with low electricity tariffs. On the one hand the low electricity tariffs could not be additionally decreased by competition and on the other hand with low electricity tariffs one can hardly expect private investments. The examples of the countries are Brazil and Russia.
- *Political motives*, including external factors (joining WTO, requirements of the International Monetary Fund, the World Bank, the European Union and others). These motives and factors influenced the situation in the countries of South America, Europe, and Russia.

  The whole set of these and probably other reasons can explain such a strange phenomenon as deregulation of electric power industry.

3. *Flaws and consequences of competitive market can be eliminated only by restoring state regulation in electric power industry.* This refers to regulation of electricity prices and centralized planning of EPS expansion. Without regulation price rise and generation capacity shortage are inevitable. This should be expected both in the successful-so-far competitive markets in Western Europe and in the markets in Russia.

9.5 Directions for Further Studies

The present book can hardly pretend to the completeness of studies on the phenomenon of power industry deregulation. Many issues are either superficially considered or insufficiently substantiated. In the author's opinion further studies should concern the following problems and points.

1. The methodology or system of state regulation of electricity market. Here it is important to study the experience of regulation in China, the states of the USA that retained regulation, Japan, France, and Brazil.
2. A more thorough analysis of possible deregulation results in Western Europe, the USA, and Australia, in the context of the world financial crisis.
3. The study of reliability problems in transition to competitive electricity markets.
4. The problems of electricity market development in Russia:

   – Investment sources and mechanisms for new power plants and transmissions.
   – Special features, flaws and possible results of NOREM operation.
   – Specific features of territorial generation companies participation in electricity market including interrelation and efficiency of combined heat and electricity production at co-generation power plants under regulated heat energy prices, possible monopolism of TGC, etc.
   – Most expedient transformation of the market in the event that the state regulation of electricity prices is restored.

Appendix A
Derivation of Expressions for the Investment Component of Electricity Price (Tariff)[1]

Construction of power plants raises an electricity price or tariff (regulated price) by the value of the investment component r against the costs of operating power plants. Here we present the derivation of mathematical expressions for this component for three mechanisms of financing new power plants. For the sake of greater vividness, taxes are neglected and some simplifications that are explained in the text are made.

A.1 Competitive Market (Mechanism 3 of Financing)

In the competitive market construction of power plants is financed by a private investor and the investments should be paid back over a period of T_R years with an annual interest σ owing to sale of electricity produced by this power plant.

Let us introduce the following notations:

N—installed capacity of the power plant, kW
k—specific investments, $/kW
h—annual number of installed capacity utilization hours, h/year

We will not take into account the period of power plant construction and assume that the investments

$$K = kN$$

are made by the beginning of the first year of its operation (if the construction period is taken into consideration, the value of K will increase because of investment "freeze").

Assume that the investments and the accrued interest are repaid at the end of each year t during T_R years by equal parts ΔD ($D = \Delta D T_R$ – Total sum of payments). Then if we succeed in determining ΔD, the investment component of electricity

[1] Derivation of formulas for the investment component is given in accordance with Appendix 1 in [19].

DOI 10.1007/978-1-4419-5612-5,

price r_3 in the competitive market can be calculated by dividing ΔD by the annual output of the power plant (index "3" denotes Mechanism 3 of financing):

$$r_3 = \frac{\Delta D}{hN}. \tag{A.1}$$

The annual payments ΔD can be determined by sequential calculation of the debt B_t remaining at the end of the year t after charge of the annual interest σ on the debt of the previous year and the payment ΔD:

$$B_t = (1+\sigma)B_{t-1} - \Delta D \tag{A.2}$$

for all T_R years. In the last year $t=T_R$ the debt will completely be repaid and $B_{T_R} = 0$.

Consider the process of investment recovery. At the beginning of the first year of power plant operation it is natural that $B_0=K$. At the end of the first year the debt will be:

$$B_1 = (1+\sigma)K - \Delta D.$$

At the end of the second year it will make up:

$$B_2 = (1+\sigma)B_1 - \Delta D = (1+\sigma)^2 K - \Delta D[(1+\sigma)+1].$$

At the end of the third year it will be equal to:

$$B_3 = (1+\sigma)B_2 - \Delta D = (1+\sigma)^3 K - \Delta D[(1+\sigma)^2 + (1+\sigma)+1].$$

At the end of the next t-th year we will have:

$$B_t = (1+\sigma)^t K - \Delta D[(1+\sigma)^{t-1} + (1+\sigma)^{t-2} + \cdots + (1+\sigma)+1].$$

Finally, for the last year $t=T_R$, when the debt is completely repaid, we will obtain the expression:

$$\begin{aligned} B_{T_R} &= (1+\sigma)^{T_R} K - \Delta D \\ &\times [(1+\sigma)^{T_R-1} + (1+\sigma)^{T_R-2} + \cdots + (1+\sigma)+1]. \end{aligned} \tag{A.3}$$

It is known (it can be shown) that the expression (sum) in square brackets can be transformed to the form:

$$[\cdots] = \frac{(1+\sigma)^{T_R} - 1}{\sigma} \tag{A.4}$$

Then in terms of $B_{T_R} = 0$ the following expression for the annual payments ΔD will be obtained:

$$\Delta D = K\frac{\sigma}{1-(1+\sigma)^{-T_R}} = K(\text{CRF}), \tag{A.5}$$

where CRF is a capital recovery factor (see, for example, [23]):

$$\mathrm{CRF} = \frac{\sigma}{1-(1+\sigma)^{-T_{\mathrm{R}}}} = \frac{\sigma(1+\sigma)^{T_{\mathrm{R}}}}{(1+\sigma)^{T_{\mathrm{R}}}-1}. \tag{A.6}$$

Substituting (A.5) into (A.1), we will obtain an expression for the investment component in the competitive market:

$$r_3 = \frac{K}{hN}\mathrm{CRF} = \frac{k}{h}\mathrm{CRF}. \tag{A.7}$$

At $T_{\mathrm{R}}=10$ years and $\sigma=0.1$, CRF $= 0.163$ and the investment component will be equal to:

$$r_3 = 0.163\frac{k}{h}.$$

If the investments were recovered in 10 years without the interest ($\sigma=0$), the investment component would equal:

$$r_3 = 0.1\frac{k}{h}.$$

Hence, charge of the interest rate $\sigma=0.1$ (during 10 years) increases the investment component by 1.63 times.

A.2 Regulated Monopoly with Self-Financing (Mechanism 1 of Financing)

In the regulated monopoly investments in generation capacity expansion are allocated to the entire output of all operating power plants in EPS. Therefore, it is necessary to consider construction of all power plants or general expansion of EPS generation capacities.

Let the system installed capacity increase (following electricity consumption) at an annual pace of expansion λ:

$$N_t = N_0(1+\lambda)^t. \tag{A.8}$$

Assume the following simplifications:

- EPS consists of single-type power plants (operating and new) with the invariable specific investments k and the number of utilization hours h.
- Power plants are built for one year and commissioned at the end of the year.
- A discrete character of power plant capacities will not be taken into account when considering only the required capacity increase $\Delta N_t = N_t - N_{t-1}$.
- Investments in power plants only will be considered (expansion of electric networks will call for an additional rise of the investment tariff component that is not dealt with here).

- Taxes, auxiliaries of power plants, and losses in electric networks will be neglected.

Under these assumptions the generation capacity increase to be provided at the expense of the investment component of tariff will make up:

$$\Delta N_t = N_t - N_{t-1} = N_0(1+\lambda)^t - N_0(1+\lambda)^{t-1}$$
$$= N_0(1+\lambda)^{t-1}(1+\lambda-1) = \lambda N_{t-1}. \qquad \text{(A.9)}$$

The capital investments required to provide this increase equal:

$$K_t = k\Delta N_t = k\lambda N_{t-1}. \qquad \text{(A.10)}$$

These capital investments are allocated among the outputs of power plants available by the end of the previous year $Q_t = hN_{t-1}$, and the investment component of tariff will equal (index "1" denotes financing Mechanism 1):

$$r_1 = \frac{K_t}{Q_t} = \frac{k\lambda N_{t-1}}{hN_{t-1}} = \lambda\frac{k}{h}. \qquad \text{(A.11)}$$

Comparison of this expression with formula (A.7) for the investment component under the competitive market shows that here instead of CRF there is a pace λ that is normally much lower. For example, for Russia now we can suppose $\lambda = 0.05 - 0.08$, while at $T_R = 10$ years and $\sigma = 0.1$, CRF$=0.163$.

A.3 Regulated Monopoly Borrowing Credits and Single-Buyer Market (Financing Mechanism 2)

In this case new power plants are constructed at the expense of borrowings (in the regulated monopoly) or by private investors (under the single-buyer market). The investment component of tariff includes the repayment of credits or private investments during some period T_R at an interest on capital σ. The application of the same financing mechanism is conditioned by two factors:

1. The repayment of credits and investments is *guaranteed* (there is no financial risk), i.e., the interest rate σ can be relatively low and approximately equal.
2. The amount of credits or investments to be repaid is allocated among the outputs of *all* operating power plants in EPS (as in Mechanism 1). Therefore, the investment component r_2 in Mechanism 2 will depend on a pace of expansion λ.

The formula for r_2 will be derived below for the regulated monopoly borrowing credits, which is simpler and more illustrative. For the single-buyer market this formula will be identical.

Suppose that in all years the credits are taken at one and the same interest rate σ and have to be repaid by equal annual amounts in an identical period T_R. Let all simplifications made in the previous section remain valid. Then the following expressions will be valid: (A.8) for the installed capacity of system N_t; (A.9) for the capacity increase ΔN_t; (A.10) for annual capital investments K_t.

At the same time there will be considerable changes as compared to self-financing. The repayment of credit in the amount of K_t, taken in the year t to commission capacities ΔN_t, will start only next year $t+1$. In the year t it is necessary to allocate the repayments of credits taken in the previous years

$$t-1, t-2, \ldots, t-T_{\mathrm{R}}$$

against the capital investments

$$K_{t-1}, K_{t-2}, \ldots, K_{t-T_{\mathrm{R}}},$$

to the output of this year $Q_t = hN_{t-1}$.

The capital investments are calculated by expression (A.10).

As is shown in Sect. A.1, when repaying credits (or investments) by equal parts in T_{R} years, the annual repayments of previously taken credits, according to expression (A.5), will be equal to:

$$K_{t-1} \cdot \mathrm{CRF},\ K_{t-2} \cdot \mathrm{CRF}, \ldots, K_{t-T_R} \cdot \mathrm{CRF}.$$

Hence, taking into account expressions (A.8) and (A.10) in the year t the credits to be repaid equal the amount:

$$D_t = \mathrm{CRF} \cdot k\lambda\, [N_{t-2} + N_{t-3} + \cdots + N_{t-T_{\mathrm{R}}} + N_{t-T_{\mathrm{R}}-1}]_1. \tag{A.12}$$

Transform the sum in square brackets:

$$\begin{aligned} [\cdots]_1 &= N_0\, [(1+\lambda)^{t-2} + (1+\lambda)^{t-3} + \cdots + (1+\lambda)^{t-T_{\mathrm{R}}} + (1+\lambda)^{t-T_{\mathrm{R}}-1}] \\ &= N_0(1+\lambda)^{t-T_{\mathrm{R}}-1} \left[(1+\lambda)^{T_{\mathrm{R}}-1} + (1+\lambda)^{T_{\mathrm{R}}-2} + \cdots + (1+\lambda) + 1\right]_2. \end{aligned} \tag{A.13}$$

Here, it can be shown on analogy with expressions (A.3) and (A.4) that the sum in square brackets amounts to:

$$[\cdots]_2 = \frac{(1+\lambda)^{T_{\mathrm{R}}} - 1}{\lambda}. \tag{A.14}$$

By substituting (A.14) into (A.13), we will have:

$$\begin{aligned} [\cdots]_1 &= N_0 \frac{(1+\lambda)^{t-T_{\mathrm{R}}-1}[(1+\lambda)^{T_{\mathrm{R}}} - 1]}{\lambda} \\ &= N_0 \frac{(1+\lambda)^{t-1}[1-(1+\lambda)^{T_{\mathrm{R}}}]}{\lambda} = N_{t-1} \frac{1-(1+\lambda)^{-T_{\mathrm{R}}}}{\lambda}. \end{aligned} \tag{A.15}$$

Now, by substituting (A.15) into (A.12) we will obtain:

$$D_t = \mathrm{CRF} \cdot kN_{t-1}[1-(1+\lambda)^{-T_{\mathrm{R}}}]. \tag{A.16}$$

By dividing D_t by an annual output of electricity $Q_t = hN_{t-1}$, we will find the investment component of tariff for the monopoly borrowing credits (and for the single-buyer market):

$$\begin{aligned} r_2 &= \frac{D_t}{Q_t} = \mathrm{CRF}\frac{k}{h}\left[1-(1+\lambda)^{-T_\mathrm{R}}\right] \\ &= \frac{k}{h}\frac{\sigma}{1-(1+\sigma)^{-T_\mathrm{R}}}\left[1-(1+\lambda)^{-T_\mathrm{R}}\right]. \end{aligned} \tag{A.17}$$

As is seen under Mechanism 2, the expression for the investment component of tariff has appeared to be most complicated, since it depends on T_R, σ, and λ.

References

1. On Restructuring the Electric Power Industry in the Russian Federation/Decree of the Government of the Russian Federation of July 11, 2001, no. 526
2. Sioshansi F (2006) Electricity Market Reform: What Have We Learned? What Have We Gained? Electr J, vol. 19, no. 9:70–83
3. Coppens F, Vivet D (2004) Liberalization of Network Industries: Is Electricity an Exception to the Rule? National Bank of Belgium. Working Paper no. 59, September. http://www.nbb.be
4. Woo C, Lloyd D, Tishler A (2003) Electricity Market Reform Failures: UK, Norway, Alberta and California. Energy Policy, vol. 31:1103–1115
5. Banks F (2002) A Simple Economic Analysis of Electricity Deregulation Failure. OPEC Rev, June:169–181
6. Joskow P (2003) The Difficult Transition to Competitive Electricity Market in the U.S. Proceeding of the Conference "Electricity Deregulation: Where From Here?" Texas A&M University, April 4
7. Apt J (2005) Competition Has Not Lowered US Industrial Electricity Prices. Carnegie Mellon Electricity Industry Center Working Paper CEIC-05-01. http://www.cmu.Edu/electricity
8. Rudnick H, Barroso L, Skerk C, Blanko A (2005) South American Reform Lessons. IEEE Power Energy Mag, vol. 3, no. 4:49–59
9. Woo C, King M, Tishler A, Chow L (2006) Cost of Electricity Deregulation. Energy, vol. 31:747–768
10. Tishler A, Woo C (2006) Likely Failure of Electricity Deregulation: Explanation with Application to Israel. Energy, vol. 31:845–856
11. U.S. Electricity Rates Remain Highest in Deregulated States. Bus Dev Transm Distribution World, August 2006, vol. 58, no. 8
12. McCullough R, Howard B, Deen M (2008) The High Cost of Restructuring. Public Util Fortn, February:54–58. http://www:fortnightly.com
13. Belyaev L (1968) Optimal Control of Power Systems with Hydro Power Plants, Using Probabilistic Methods: Summary of the Doctoral Theses. SB RAS, Novosibirsk (in Russian)
14. Belyaev L, Zeiliger A (eds) Problems of Constructing Automated Information Control Systems of Power System Expansion. Issue 2. Structure and Construction Principles of Phase 1 of AICS (1975), SEI SB AS of the USSR, Irkutsk (in Russian)
15. Belyaev L (1976) A Systems Approach to Development Planning of the Fuel Power Industry of a Planned Economy Country. RR-76-6. International Institute for Applied Systems Analysis, Laxenburg, Austria
16. Belyaev L, Voitsekhovskaya G, Saveliev V et al (1980) System Approach to the Management of Power Industry Development. Nauka, Novosibirsk (in Russian)
17. Belyaev L (2007) Electricity Markets: Comparing Competitive and Single Buyer Markets. IEEE Power Energy Mag, vol. 5, no. 3:16, 18, 20, 22, 24–26

L. S. Belyaev, *Electricity Market Reforms*,
DOI 10.1007/978-1-4419-5612-5, © Springer Science+Business Media, LLC 2011

18. Belyaev L (2005) The Need for a Regulation of the Generation Capacity Development in a Competitive Electricity Market. Proceed Int Conf "2005 IEEE St Peter Power Tech," St Petersburg, Russia
19. Belyaev L, Podkovalnikov S (2004) Electricity Market: Problems in Expansion of Generation Capacities. Nauka, Novosibirsk (in Russian)
20. Belyaev L (2004) Marginal Prices in the Wholesale Electricity Market and "Producer's Surplus." Energiya: ekonom, tekhnika, ekologia, no. 4:2–9 (in Russian)
21. Belyaev L (2007) Does Russia Need a Competitive Market in Electric Power Industry? Energiya: ekonom, tekhnika, ekologia, no. 4:2–7 (in Russian)
22. Belyaev L, Khudyakov V (2008) Worldwide Experience in Electricity Market Reform. Northeast Asia Energy Forum, vol. 5, no. 4:32–53. http://www.neasiaenergy.net
23. Stoft S (2002) Power Systems Economics: Designing Markets for Electricity. John Wiley & Sons, Inc., New York
24. Belyaev L, Podkovalnikov S, Saveliev V, Chudinova LYu (2008) Efficiency of Interstate Electric Ties. Nauka, Novosibirsk (in Russian)
25. Volkenau I, Zeiliger A, Khabachev L (1981) Economics of Creating Electric Power Systems. Energiya, Moscow (in Russian)
26. Rokotyan SS, Shapiro IM (eds) (1985) Reference Book on Designing Electric Power Systems, 3rd rev. and enl. ed. Energoatomizdat, Moscow (in Russian)
27. Economico-mathematical Optimization Models of Power Industry Designing and Planning. Collected papers (1985) Krzhizhanovsky Power Engineering Institute, Moscow (in Russian)
28. Proceedings of the Institute "Energosetproekt." (1970) Issue 1. Energiya, Moscow (in Russian)
29. Designing and Operation of Power Systems and Electric Networks. Proceedings of the Institute "Energosetproekt." (1980) Issue 20. Energiya, Moscow (in Russian)
30. De Vries L (2004) Securing the Public Interest in Electricity Generation Market. Ph.D. dissertation. Department of Technology, Policy and Management, Delft University of Technology, the Netherland
31. Rozanov M (1974) Reliability of Electric Power Systems. Energiya, Moscow (in Russian)
32. Hunt S, Shuttleworth G (1996) Competition and Choice in Electricity. John Wiley & Sons, Ltd., Chichester, England
33. Hingorany N, Gyugui L (1999) Understanding FACTS: Concepts and Technology of Flexible AC Transmission System. IEEE Press, Piscataway, NJ
34. McConnel C, Brue S (1999) Economics: Principles, Problems and Policies, 14th edn. McGraw-Hill, Toronto
35. Melentiev L (1976) Optimization of Expansion and Control of Large Energy Systems. Vysshaya shkola, Moscow (in Russian)
36. Melentiev L (1983) System Studies in Energy, 2nd enl. edn. Nauka, Moscow (in Russian)
37. Belyaev L, Rudenko Yu (eds) (1986) Theoretical Grounds for System Studies in Energy. Nauka, Novosibirsk (in Russian)
38. Kirschen D, Strbac G (2004) Fundamentals of Power System Economics. John Wiley & Sons, Ltd., Chichester, England
39. On Electric Power Industry. (2003) Federal Law of the Russian Federation, March 26, no. 35-f3 (in Russian)
40. Fischer S, Dornbusch R, Schmalensee R (1988) Economics, 2nd ed. McGraw-Hill, New York
41. Haiman D (1992) Modern Microeconomics: Analysis and Applications, vol. 2. Finansy i statistika, Moscow (in Russian)
42. Xu S, Chen W (2006) The Reform of Electricity Power Sector in the PR of China. Energy Policy, vol. 34:2455–2465
43. Markovich I (1969) Operating Conditions of Energy Systems. Energiya, Moscow (in Russian)
44. Gonchukov V, Gornshtein V, Krumm L et al (1979) Automation of Power Interconnection Control. Sovalov S (ed). Energiya, Moscow (in Russian)

45. Rudenko Yu, Cheltsov M (1974) Reliability and Backup in Electric Power Systems. Nauka, Novosibirsk (in Russian)
46. Voropai N, Ershevich V, Luginsky Ya et al (1984) Control of Large Power Pools. Sovalov S (ed). Energoatomizdat, Moscow (in Russian)
47. Sovalov S, Semenov V (1988) Emergency Control in Energy Systems. Energoatomizdat, Moscow (in Russian)
48. Berkovich M, Gladyshev V, Semenov V (1991) Automatic Devices of Energy Systems. Energoatomizdat, Moscow (in Russian)
49. Volkov E, Barinov V, Manevich A (2001) Problems and Prospects of Russia's Power Industry Development. Energoatomizdat, Moscow (in Russian)
50. Podkovalnikov S (2003) Investing in Electric Power Industry in the USSR and in Russia. Energiya: ekon, tekhnika, ekologia, no. 5:12–18 (in Russian)
51. Stiglitz J (2003) The Roaring Nineties. Seeds of Destruction. W.W. Norton & Company, New York
52. Kucherov Yu, Kucherova O, Kapolyi L, Rudenko Yu (1996) Reliability and Effectiveness in Operation of Bulk Transnational EPS. Kapolyi L (ed). Nauka Sib Publ Com of RAS, Novosibirsk (in Russian)
53. Miisgenst F (2006) Quantifying Market Power in the German Wholesale Electricity Market Using a Dynamic Multi-Regional Dispatch Model. J Ind Econ, vol. 54, no. 4:471–498
54. Vasiliev Yu, Glebov I, Demirchan K et al (2001) Preconditions for Self-Sufficient Development of Russia's Power Industry. Izv RAS Energetika, no. 3:3–32 (in Russian)
55. Platonov V (2000) Goals and Ways of Energy Restructuring. IBRAE RAS, Moscow. Preprint of IBRAE RAS no. IBRAE-2001-10 (in Russian)
56. Maslennikov V (2001) How to Help Russia's Energy Out of the Crisis. Energiya: ekon, tekhnika, ekologia, no. 6:2–6 (in Russian)
57. Nakoryakov V (2003) On the Project of Russia's Power Industry Restructuring. Energeticheskaya Politika, no. 1:54–67 (in Russian)
58. Kuzovkin A (2006) Restructuring of Electric Power Industry and Energy Security. JSC "Inst of microecon," Moscow (in Russian)
59. Batenin V, Maslennikov V (2004) What Energy Perspectives Does Russia Have. Energiya: ekon, tekhnika, ekologia, no 9:2–11 (in Russian)
60. Banks F (2007) A New Lecture on Electricity Deregulation Failure. Energy Pulse, October 24. www.energypulse.net
61. Schweppe F, Caramanis M, Tabors R, Bohn R (1988) Spot Pricing of Electricity. Kluwer Academic Publishers, Boston
62. Thomas S (1996) The Development of Competition. In: Surrey J (ed) The British Electricity Experiment. Farthean Publications Ltd., London:67–94
63. Li Y, Flynn P (2004) Deregulated Power Prices: Comparison of Diurnal Patterns. Energy Policy, vol. 32:657–672
64. Zareipour H, Bhattacharya K, Canizares C (2007) Electricity Market Price Volatility: The Case of Ontario. Energy Policy, vol. 35:4739–4748
65. Fehr V, Harbord D (1995) Capacity Investment and Long Run Efficiency in Market-Based Electricity Industries. In: Olsen O (ed) Competition in the Electricity Supply Industry. Jurist-og Okonomforbundets Forlag. DJOF Publishing, Copenhagen:137–153
66. The General Scheme of Electric Power Object Placement up to 2020/Approved by the Decree of the RF Government of February 22, 2008, no. 215-p (in Russian)
67. Gornshtein V (1959) Most Beneficial Operating Conditions for Hydro Power Plants in Power Systems. Gosenergoizdat, Moscow (in Russian)
68. Gornshtein V (ed) (1981) Optimization Methods of Energy System Operation. Energoizdat, Moscow (in Russian)
69. Lessons Learnt from Liberalization of Electricity Markets. Transl from Engl (2005) International Energy Agency, Paris (in Russian)

70. Dimson E (1989) The Discount Rate for a Power Station. Energy Econ, vol. 11, no 3:175–180
71. Ibbotson R, Singuefield R (1987) Stocks, Bonds, Bills and Inflation. 1987 Yearbook. Ibbotson Associates, Chicago
72. Levy H, Sarnat M (1986) Capital Investment and Financial Decisions. Prentice Hall, Englewood Cliffs, NJ
73. Tung Au, Thomas P (1992) Engineering Economics for Capital Investment Analysis. Prentice Hall, Englewood Cliffs, NJ
74. Vilensky P, Lifshits V, Smolik S (2008) Efficiency Estimation of Investment Projects: Theory and Practice. Delo, Moscow (in Russian)
75. Bronshtein I, Semendyaev K (1981) Mathematics Reference Book. Nauka, Moscow (in Russian)
76. U.S.-Canada Power System Outage Task Force. Final Report on the August 14, 2003 Blackout in the United States and Canada: Causes and Recommendations. (2004) April. https://reports.energy.gov
77. Public Utility Commission of Texas. Investigation into April, 2006 Rolling Blackouts in the Electric Reliability Council of Texas Region. Preliminary Report. (2006) April 24
78. Bancalari E, Chan P (2005) Adaptation of the SGT6-6000G to a Dynamic Power Generation Market. POWER-GEN Int 2005. December 6–8, Las Vegas, NV
79. State of the Markets Report. Federal Energy Regulatory Commission. Office of Mark Overs and Investigations, June 2005
80. Thomas S (2004) The Ontario Government's Proposals on Electricity Restructuring: Comments by Public Service International Research Unit/PSIRU. www.psiru.org
81. Wu F, Fu S (2005) China's Future in Electric Energy. IEEE Power Energy Mag, vol. 3, no. 4, July/August:32–38
82. Yadav R, Roy A, Khaparde S, Pentayya P (2005) India's Fast Growing Power Sector. IEEE Power Energy Mag, vol. 3, no. 4, July/August:39–48
83. Bhushan B (2006) A Market Design for Developing Countries. CIGRE Paper no. C5-208. Paris
84. Gjerde O, Fosso O, Bae J et al (2006) Interfaces Between System Operators, Market Operators and Other Market Actors. CIGRE Paper no. C5-216. Paris
85. Privatization of KEPCO Comes to a Halt. (2004) Biz Econ, Seoul, May 31
86. Clavel P, Bauchet S (2006) Organization of the French Electricity Sector. CIGRE Paper no. C5-106. Paris
87. Tesseron J (2008) Mission: Reliability. IEEE Power Energy Mag, vol. 6, no. 1:42–48
88. Arango S, Dyner I, Larsen E (2006) Lessons from Deregulation: Understanding Electricity Markets in South America. Utility Policy, vol. 14:196–207
89. Thomas S (2006) The British Model in Britain: Failing Slowly. Energy Policy, vol. 34, no. 5:583–600
90. MacGregor T (2001) Electricity Restructuring in Britain: Not a Model to Follow. IEEE Spectrum, June:15–16, 19
91. Swift D, Thorpe G, Baker T (2006) Interaction Between Investment Processes for Regulated Transmission and Competitive Generation in the Australian NEM. CIGRE Paper no. C5-101. Paris
92. Diakov A, Platonov V (2003) Unified Electric Power System of Russia During Market Reforms: Training Manual. Publ House of MEI, Moscow (in Russian)
93. Fundamentals of Russia's Power Industry Development Strategy till 2020. (2000) ERI RAS, Moscow (in Russian)
94. Nekrasov A, Sinyak Yu, Uzyakov M (2001) Russia's Energy: Economy and Reforms. IEF RAS, Moscow (in Russian)
95. Kutovoi G, Makarov A, Shamraev N (1997) Creation of Favorable Conditions for Development of Russia's Power Industry in a Market Environment. Teploenergetika, no. 1:2–7 (in Russian)

96. Dorofeyev V, Mikhailov V, Fraer I (2000) Market for Electrical Energy and Capacity in Russia: What Will It Be? Edelman V (ed). Energoatomizdat, Moscow (in Russian)
97. On specific features of power industry operation in the transitional period and on the introduction of amendments to certain legislative acts of the Russian Federation and recognition of some legislative acts of the Russian Federation as expired due to adoption of the Federal Law "On electric power industry" (2003) Fed Law of the Russian Federation of March 26, no. 36–Ф3 (in Russian)
98. On Improvement of the Order of the Wholesale Electricity (Capacity) Market Operation. (2006) Decree of the RF Government of August 31, no. 529 (in Russian)
99. Mikhailov V (2004) Russian Electricity Market: Transitional Period. Polygraph Center of MEI, Moscow (in Russian)
100. Tukenov A (2005) Electricity Market: From Monopoly to Competition. Energoatomizdat, Moscow (in Russian)
101. Ayuev B (2007) Electricity Markets and Their Implementation in Russia's UPS. UrB RAS, Ekaterinburg (in Russian)
102. Voropai N, Ivanova E, Trufanov V (2007) Power Industry Development Problems, Methods and Mechanisms for Their Solution in a Market Environment. IEF RAS, Moscow (in Russian)
103. Maximov B, Molodyuk V (2008) Theoretical and Practical Foundations of Electricity Market: Study Guide. Publ House of MEI, Moscow (in Russian)
104. Platonov V (2006) Analysis of Development Strategy and Problems of Restructuring in Russia's Electric Power Industry. NPI, Novocherkassk (in Russian)
105. Lvov D, Batenin V, Kudryavy V et al (2007) Loss of State Control Over Power Market Will Lead to Economic Breakdown. Energo-Info, no. 4:44–46 (in Russian)
106. Mastepanov A (2001) Russia's Fuel and Energy Complex at the Turn of a Century—Status, Problems, and Prospects of Development (Information and Analytical Review). Publ House "Sovremennye tetradi," Moscow (in Russian)
107. Makarov A (2007) Russia's Electric Power Industry till 2030: Outlines of the Desirable Future. ERI RAS, Moscow (in Russian)

Index

L. S. Belyaev, *Electricity Market Reforms*,
DOI 10.1007/978-1-4419-5612-5, © Springer Science+Business Media, LLC 2011

The manufacturer's authorised representative in the EU is Springer Nature Customer Service Centre GmbH, Europaplatz 3, 69115 Heidelberg, Germany. If you have any concerns regarding our products, please contact ProductSafety@springernature.com

Printed and bound by CPI Group (UK) Ltd, Croydon, CR0 4YY
10/10/2026
02231227-0004